Wireless Foresight

se
i
)
3 i

2006

nditions

Wireless Foresight

Scenarios of the mobile world in 2015

Bo Karlson
Aurelian Bria
Jonas Lind
Peter Lönnqvist
Cristian Norlin

All of Wireless@KTH, Sweden

WILEY

Other Wiley Editorial Offices

John Wiley & Sons Inc., 111 River Street, Hoboken, NJ 07030, USA

Jossey-Bass, 989 Market Street, San Francisco, CA 94103-1741, USA

Wiley-VCH Verlag GmbH, Boschstr. 12, D-69469 Weinheim, Germany

John Wiley & Sons Australia Ltd, 33 Park Road, Milton, Queensland 4064, Australia

John Wiley & Sons (Asia) Pte Ltd, 2 Clementi Loop #02-01, Jin Xing Distripark,
Singapore 129809

John Wiley & Sons Canada Ltd, 22 Worcester Road, Etobicoke, Ontario, Canada
M9W IL1

Wiley also publishes its books in a variety of electronic formats. Some of the content
that appears in print may not be available in electronic books.

Library of Congress Cataloging-in-Publication Data

Wireless foresight : scenarios of the mobile world in 2015 / Bo Karlson . . . [et al.].
    p. cm.
    Includes bibliographical references and index.
    ISBN 0-470-85815-X (alk. paper)
    1. Wireless communication systems—Forecasting.   2. Mobile communication
systems—Forecasting.   3. Telecommunication—Forecasting.   I. Karlson, Bo.
  TK5103.2.W57325 2003
  384.5′01′12—dc21
                                         2003053544

British Library Cataloguing in Publication Data

A catalogue record for this book is available from the British Library

ISBN 0-470-85815-X

Typeset in 10/12pt Times NR by Graphicraft Limited, Hong Kong
Printed and bound in Great Britain by Biddles Ltd, Guildford and King's Lynn
This book is printed on acid-free paper responsibly manufactured from sustainable
forestry in which at least two trees are planted for each one used for paper
production.

Contents

Preface

This is a book about the future. It deals with the development of the wireless communications industry and technology during the coming 10–15 years. This is indeed an exciting field with a great potential for changing society and the way we live and work. The communications industry is clearly one of the largest and most important industries in the world. Markets and companies are truly global and millions of people get their livelihood from working as operators, vendors, service providers, application developers, content owners, and suppliers of components, products, and services. We have seen an absolutely amazing growth in the use of information and communication technologies over the past decades. The explosive development of the internet is perhaps the most striking event during the latter half of the twentieth century. The development of wireless technology and the very rapid growth in use of mobile telephony are almost as important. Today there are more mobile telephone subscriptions in the world than there are fixed lines. A short time ago, the number of mobile phone users passed 1 billion. The internet and the mobile phone are now integral parts of our lives. It seems as if human beings have an insatiable need to communicate with each other. Perhaps communication is one of the things that make us human.

We have arrived at a moment when the internet and the mobile phone are merging. An integrated industry of IT, media, and telecommunications is gradually taking shape. New constellations of companies are being formed. New players are entering the scene and novel ways to do business are emerging, along with previously unseen value chains. At the same time, these industries currently face a serious crisis, especially the wireless communications companies. We are on the verge of introducing third-generation cellular systems but we are astonished by the cost involved in acquiring licenses, designing, building, and maintaining these highly complex networks. As with any new generation of technology, people wonder how we might use it, what services will catch on, and how much we'll be willing to pay. Many people also worry about the effects of radiation from terminals and networks.

It seems we are at a crossroads. This is therefore an excellent time to stop for a little while, take a step back, and reflect on development in the past, what happens in the present and, above all, think about the future. This is precisely the purpose of our book. We ask you to spend a few hours reading the scenarios of the wireless world in 2015. We don't think, and perhaps we don't even hope, that you will draw the same conclusions as we have. You might not even agree with any of the scenarios. But that doesn't matter. If our scenarios can help you unlock your mind and inspire you to create your own vision of the future, then we have succeeded.

In a world of immense complexity such as ours, it is impossible to understand every detail and every aspect of the present and the future. This makes it very difficult to produce good forecasts or predictions and even harder to convince people they are accurate. Our way is different. The four scenarios in this book are not intended as predictions but as more holistic images of the future. They point in four different directions and we don't consider any scenario more likely to happen than the others. As individuals we might have our personal favorite but that doesn't make it more likely to occur. Perhaps you will find your own favorite.

We hope that we can inspire you to reflect on the future of wireless communications and that this will make you better prepared for what the future might hold. The scenarios were developed at Wireless@KTH, a center for research and education on wireless systems created by the Royal Institute of Technology (KTH) in Stockholm in cooperation with industry.

It has indeed been fun to write the scenarios and then reflect on them, but also complex and challenging. We are extremely grateful to many people who contributed in various ways. First and foremost, we express our gratitude to Professor Jens Zander, who came up with the idea of developing scenarios as a way of preparing for the future. With his expertise, personal network, patience, and good mood, he has been a wonderful support and a source of inspiration. We are also much indebted to Claes Beckman, the former director of Wireless@KTH, who with his amazing enthusiasm and knowledge has helped us immensely. We also express our gratitude to Anna van Bunningen and Sofia Lentell from the Boston Consulting Group (BCG), who participated in the early phases of the scenario work. Their contribution was invaluable in creating efficient work procedures, driving discussions, and organizing brainstorming sessions. Their expertise as consultants and their knowledge of the wireless industry were also very valuable. We extend a special thank-you to Professor Gerald 'Chip' Maguire for the detailed feedback and provocative ideas he has given us.

We are also indebted to many other experts from industry and academia who have provided us with much support, encouragement, and feedback: Curt Andersson (Europolitan), Sujata Bhattacharya, Cay Bond (Promostyl), Soki Choi (Blue Factory), Bo Dahlbom (SITI), David Dean (BCG), Bernt Ericson (Ericsson), Håkan Eriksson (Ericsson), Göran Eriksson (Telia), Kerstin Frenning (Telia), Fredrik Gessler (Ericsson), Tim Giles (KTH), Jacob Gramenius (3), Per Hallius (BCG), Bo Harald (Nordea), Ulf J. Johansson (Europolitan), Ulf Jonströmer (Brainheart), Gunnar Karlsson (KTH), Anna Kelly (Private Gateway), Stefan Krook (Glocalnet), Erik Kruse (Ericsson), Vicky Long, Magnus Madfors (Ericsson), Johan Montelius (KTH), Östen Mäkitalo (Telia), Bengt Nordström (Northstream), Zeth Nyström (3), Björn Ottersten (KTH), Mats Palmblad (BCG), Harald Pobloth (KTH), Bi Puranen (Bikupan), Per-Gunnar Sidén (Pegasus Affärsutveckling), Ben Slimane (KTH), Per Stein (Intel), Bertil Thorngren (Stockholm School of Economics), and Annika Waern (Game Federation). We have also received wonderful support from the staff at Wireless@KTH: Lise-Lotte Wahlberg, Anna Almlöw, Cecilia Forssman, and Linn Wahlberg.

All remaining errors and inconsistencies, our opinions, and the overall view of industry development are of course only our own responsibility. We are extremely grateful to all of you.

Bo Karlson
Aurelian Bria
Jonas Lind
Peter Lönnqvist
Cristian Norlin
Stockholm, March 2003

1

Introduction

The Wireless Industry at a Crossroads

Telecommunications is a global business of enormous proportions. It is a $1000 billion market, making it one of the biggest business opportunities in the world, second only to sex, drugs, and arms. The wireless communications sector is also huge and growing. The global number of mobile telephone users is now well over 1 billion. In 2001 the number of wireless subscriptions passed fixed-line subscriptions. Currently about 20 billion SMS messages are sent every day, even though this service has not really taken off in one of the largest and most important markets, the US. Over 400 million mobile phones were sold in 2002. Add to this 15–20 million PDAs (personal digital assistants) and 25–30 million laptop computers, many of which have wireless communication capabilities (Morgan Stanley 2001). The first (analog) cellular systems were introduced on the mass market only about 20 years ago, in the early and mid 1980s and the dominating second-generation digital standard, GSM, was introduced in 1992. Still, as late as 1997 the global number of mobile phone subscriptions was less than 150 million (Lindmark 2002; Economist 2001). The mobile phone is becoming an extension of our body. The development during the last decade has been truly amazing.

The telecommunications market can be divided into end users, equipment and terminal vendors, operators, service providers, service and application developers, and subsuppliers. The annual turnover of the industry supplying wireless equipment and terminals can be estimated at well over $100 billion worldwide, of which about $40 billion is revenue from sales of mobile infrastructure. Wireless telecommunications is indeed a very important sector and a truly global business, with markets spanning the world and large multinationals as well as small companies competing fiercely. For individual nations, such as Sweden, the telecom sector is even more important, currently accounting for about 8% of the country's total exports.

At the same time, the telecommunications industry is facing severe difficulties. For the first time, growth rates for mobile telephone subscriptions have slowed down in many important markets. Blinded by the hype in the late 1990s and early 2000s, operators have spent enormous amounts on licenses for third-generation cellular systems (3G systems); in Western Europe alone they come to around $120 billion. The investments for building these systems are expected to be at least of the same size. Add to that possible subsidy of terminals, as in GSM, of the same magnitude. In most countries, especially in Europe, 3G is delayed compared to the original plans. 3G services will probably not be launched on a broad scale until 2004 at the earliest.

Despite the tremendous growth during the 1990s and the current problems, the long-term prospects are promising. In 2003 half the world has yet to make a phone call and global penetration of cellular phones is still only around 20% (Deutsche Bank 2001). Even though many operators are burdened by huge debt from the astronomic fees paid for 3G licenses and low profitability on wire-line operations, the cellular operations continue to grow while still very profitable.

With the introduction of the packet-switched 2.5G and 3G cellular systems, a whole new range of mobile data services are possible, for example MMS (multimedia messaging), streaming video, teleconferencing, various types of location-based services, downloading of customized software, and speech recognition. Other types of systems for example WLANs (wireless local area networks) providing advanced services in specific locations, sometimes called hotspots, will complement the cellular systems. This has already started. The global WLAN equipment market grew by about 70% in 2002 with revenues reaching $2.1 billion (Northstream 2002). Many companies use WLANs as an integrated part of their office IT infrastructure. WLAN services in hotspots are becoming increasingly common in public locations as well, for example at airports, train stations, hotels, university campuses, and cafes (Lind 2002). Already several operators exploiting the WLAN business have entered the scene, both new and specialized so-called WISPs (wireless internet service providers) and incumbent cellular operators. According to Gartner Dataquest, a continued increase in revenues from WLAN equipment of about 20% is expected for 2003, while prices will drop by 25%. Although dwarfed by the volumes in the market for cellular products, the development is very fast, with the US as the leading market.

The introduction of wireless data services and new types of network will no doubt lead to the emergence of new players on the wireless scene and probably a restructuring of the whole industry as a result. Service and content providers, application developers, virtual operators, and wireless portals are a few examples of players who will try to capture the market using new business models to compete with incumbent operators. The merging of telecommunication, data communication, and media into an integrated industry will mean business opportunities for existing actors but also for new competitors. The competition between traditional telcos and datacom companies, on both the terminal and equipment markets, is becoming fiercer every day. Telecom standards and protocols fight for dominance with datacom standards and protocols. Alliances are being formed between companies from these two different worlds.

Even though it can be argued that communication lies at the core of being human and that we have an insatiable need to communicate with each other, it is not at all clear what lies ahead for the communications industry. Will there really be a demand for all those services that are possible with the introduction of third-generation cellular systems and WLANs in hotspots? How much are we willing to pay for these services and at what price can they be offered? The difficulties of providing a broad range of data services over a variety of networks should not be underestimated. Going from the relatively simple world of providing voice and simple messaging services to the far more complex wireless world of the future is certainly not simple. Can the engineering challenges facing infrastructure vendors be solved so that user demands can be met? Will new terminal technology be developed so that the current problems with cumbersome human/machine interfaces (input/output units, small screens, etc.) are solved? Will new value chains and business models emerge so that all players necessary to create this new wireless world find it worthwhile to participate?

It seems the telecommunications industry in general and the wireless sector in particular are at a crossroads. The coming few years will indeed be very exciting.

Be Prepared for 2015

This book, however, tries to look further ahead, into the wireless world beyond 3G into 2015. Which are the most important trends in the wireless industry and what are the long-term fundamental drivers of development? What services will be used in 2015? Who will be the users and what demands will they make? Which regions and nations will lead the development? What technological and other problems have to be addressed in order to realize a positive wireless future? Which are the most important areas of research? All these questions are addressed by this book.

The book's purpose can be stated in two words: *be prepared*. Our ambition is to provide a different way of thinking when preparing for the future. As we all know, there are numerous examples of forecasts completely missing the target. Just remember the predictions made on usage of wireless services, Wap, industry growth, etc., when the telecommunication hype was at its peak in 2000. As has been stated by several people before, the only thing we can say with certainty about the future is that we don't know what will happen. But we are not arguing that traditional forecasts and predictions are bad and that they should be abolished. What we suggest is that this way of preparing for the future is not enough. Our path is different. By devising scenarios that are broad and represent reasonably holistic images of the future, a different basis for reflection and discussion is created, enabling us to better prepare for what the future might hold.

Scenarios of the Wireless World in 2015

The core of this book consists of four scenarios that describe possible wireless worlds in 2015. They are focused on the development of the wireless industry in a broad context. The scenarios are concrete images, including descriptions of the development of the telecommunications industry, the wireless systems of 2015, how these systems are used, and who are the most important users.

In "Wireless Explosion—Creative Destruction" wireless services and technology develop very rapidly, transforming industry so that the old market leaders, the traditional telcos and their equipment vendors, lose their dominant positions. The old world with closed and vertically integrated systems gives way to layered and open architectures based on the Internet Protocol (IP). Even though the previous market leaders don't vanish, it is the datacom industry that wins the market battle.

In "Slow Motion" the world moves into an economic recession following the bursting of the great telecom bubble in the middle of the 2000s. On top of that, research shows that electromagnetic radiation from mobile devices is harmful, forcing industry to retreat but finally to refocus on new and more harmless technologies. The problems with guaranteeing security and integrity in transmissions prove very difficult to solve. Users are reluctant to use wireless technology, severely affecting the industry. In 2015 the wireless industry has only just started to get back on track. We have seen a positive development in several of the big NICs (newly industrialized countries) and they are gradually catching up with the industrialized world.

"Rediscovering Harmony" involves a significant lifestyle shift in the industrialized world. Balance in life and human and environmental needs are in focus, affecting all sectors of society and industry. The migration flow into the large polluted cities ended in favor of quality of life and less stress in smaller local communities. There is large diversity in lifestyle between different groups, or tribes, in society. People live locally but think globally. There are fewer wireless services than expected around the turn of the century, but still a substantial market. Communication between people is a success but appetite for new cool and funky applications is lower than expected. The big difficulty for the wireless industry has been to rethink business models and services in relation to these changed market conditions.

Through consolidation and mergers, a few large companies have come to dominate the wireless world in "Big Moguls and Snoopy Governments." These moguls have expanded outside their original business segments. Together with the world's governments, they exert substantial control over information flow and the communication and media industries. The purpose of the governments is to protect society and individuals from cybercrime and terrorism, and to protect content owners from illegal copying. Most problems concerning security on the internet have been solved, but at the price of slow industry development. Some services are not introduced since they are illegal.

The scenarios are not predictions but possible and, hopefully, plausible descriptions of the future. They are intended as a source of inspiration when thinking about the future of wireless technology and industry. It is no coincidence that the title for the book is *Wireless Foresight*, not *Wireless Forecast*.

The scenarios are qualitative in nature. You will find no statistics, graphs, or growth figures in them. Since they are not intended as predictions, the choice was made early on to take a descriptive approach. This does not mean that we have avoided statistical material or are unaware of it. There is a lot out there, most of it compiled by authors more competent statisticians than us. The scenarios are implicitly based on various predictions, trends, and forecasts of a more quantitative nature.

Since the objective is to develop plausible and consistent descriptions of possible wireless worlds in 2015, it is necessary to have a long-term perspective and a broad scope. The temporal perspective is until 2015 and the scope is broader than the industrialized world. The development in the leading regions of today—Western Europe, North America, and parts of Asia-Pacific—is very important. Also of interest are other countries with the potential to become not only large markets but also homes to important players in the wireless industry.

Challenges for the Future

The futures described in the scenarios contain positive and negative aspects for industry as well as users, highlighting obstacles that need to be overcome and possible paths towards a positive wireless future. This makes the scenarios an excellent platform for discussing challenges facing various actors in the wireless arena, now and in the long run. The scenarios are therefore followed by a discussion of long- and short-term challenges for industry, for some especially important regions and nations, and for the research community. The images of the future described in the scenarios contain certain assumptions about the characteristics of wireless technology in 2015. These are further elaborated in the discussion on research challenges.

Creating Scenarios

The scope of the scenarios is broad. They are based on several fundamental drivers or mega-trends in technology, society, business and industry, and among users. These are things that we feel certain will be valid in 2015 as well as today. We thus paint a broad picture of the development in the larger context of the wireless industry. From these fundamental drivers, a set of more concrete trends are identified; these are 14 trends whose direction and rate of change are uncertain. Each trend is assigned different values; for example, strong or weak, fast or slow, problem solved or problem unsolved. In this way a scenario space is created and the four scenarios are defined by combining the values of the trends in different ways. Input to the scenarios comes from various sources; the most important sources are external experts from different fields in industry and academia, articles and literature on the development of the telecommunications industry, and other scenarios.

The scenarios were developed at Wireless@KTH, a center for industry-related research and education on wireless systems at the Royal Institute of Technology (KTH) in Stockholm, Sweden.

Guide to the Book

Chapters 2 to 5 describe the four scenarios: Wireless Explosion—Creative Destruction, Slow Motion, Rediscovering Harmony, and Big Moguls and Snoopy Governments. Each chapter is divided into four parts, starting with a table of the 14 trends used as dimensions when creating the scenario. The table, with different weights assigned to the trends, gives a first hint of the scenario's flavor. The second part of each scenario is a storyline in which we meet real users of wireless technology in 2015. The third part is the actual scenario describing the future wireless world and how it has evolved. Each scenario ends with a short description of the wireless technology and services used in 2015.

Chapter 6 lays out the basis for the scenarios. It describes trends, fundamental drivers (mega-trends), and theories underlying the scenarios. It is divided into three parts. The first part describes the 14 trends used to define the scenarios and gives a listing of the fundamental drivers underlying each trend. The second part contains a table with all the fundamental drivers and a short description of each driver. The chapter concludes with an overview of theoretical models underlying and supporting some of the fundamental drivers. Chapter 7 describes a number of technical implications that can be derived from the scenarios. The technical implications are formulated as statements about the wireless technology in 2015. By assuming these will be true in 2015, we also assume that the underlying research and engineering problems have been solved.

Chapter 8 discusses the main engineering challenges for creating a positive wireless world. It presents several important areas for research in order to realize a positive wireless future for industry and users. The research areas are based on the technical implications discussed in Chapter 7. Chapter 9 uses the scenarios to discuss the most important challenges facing industry in the next 10–15 years. The discussion is focused on topics we believe are critical for a positive industrial development and on problems where industry can stumble if things go wrong or are left unresolved.

Chapter 10 shifts the focus from industry to some of the most important wireless markets now and in the coming decade: the US, Europe, China, Japan, and Korea. Not only will they be important in the future as markets for wireless products and services, but also as homes to the industry's important global players. Each region has its own section in the chapter and the descriptions are rather broad, focusing on the current state of affairs and the most important trends and challenges facing the wireless industry.

Chapter 11 deals with the methodology of devising scenarios and of our work. It begins by discussing logics of creating scenarios. Then it briefly describes our methods and working procedures. The chapter ends with a brief overview of other scenarios and future-oriented studies that have been important for us as input and inspiration. Chapter 12 is a conclusion. The first section contains a full recap of the scenarios and the most important issues discussed in the book. The final section is a brief reflection on the importance and the difficulties of taking a step back from the present when thinking about the future.

Part I

Scenarios

2

Wireless Explosion—Creative Destruction

Key phrases: rapid growth, datacom winning over telecom, open IP architectures, active users, anarchistic underground culture, user-deployed networks, ad hoc networks, creative destruction, unlicensed spectrum

In this scenario, wireless applications and services are a huge success in 2015, and in a rapidly transforming industry the old market leaders have lost their dominant positions. The old telco world with closed, vertically integrated solutions gave way to layered, open architectures based on the Internet Protocol (IP) and the datacom industry won the market battle. However, in a large but maturing industry, profit margins were squeezed and the datacom market winners could never really leverage their market power.

Users were very active and drove this development towards an open IP world with skyrocketing traffic and an abundance of applications. They preferred choice over convenience and didn't accept being locked into corporate bundles. Governments released a lot of new unlicensed spectrum, undermining operator dominance and triggering a do-it-yourself wireless movement. The open-source movement, downloading of music and other uncopyrighted material, enforced these changes in consumer attitudes and the values of the underground culture gradually became mainstream. Feeling this value shift, governments were more and more reluctant to enforce restrictive IPR (intellectual property rights), further undermining profit margins.

The wireless success changed people's working habits and lifestyle. Being always connected with context-sensitive information, a growing part of the knowledge workforce could spend most of their time on the move, in meetings or traveling between meetings. Globalization continued and with it the growing trends of traveling and commuting.

Figure 2.1 shows the 14 trends defining the scenario space and how we have set their values for this scenario. The scenario begins with a personal story of life in 2015. This story is followed by the main scenario, describing important aspects of the wireless world in 2015. The chapter ends with a brief description of the technical systems and services used.

Figure 2.1 Wireless Explosion: defining dimensions. All variables are described in Chapter 6; all user segments are described in Appendix A

	Development will be more user driven	
Low usage	○ ○ ○ ○ ○ ○ ●	High (moklofs)
Low usage	○ ○ ○ ○ ○ ● ○	High (yupplots)
Low usage	○ ○ ○ ○ ● ○ ○	High (elders)
Low usage	○ ○ ○ ○ ○ ● ○	High (mobile profes.)
Low usage	○ ○ ○ ○ ○ ○ ●	High (industrial users)
	User mobility will increase	
Low travel growth	○ ○ ○ ○ ○ ○ ●	High (cars)
Low travel growth	○ ○ ○ ○ ○ ○ ●	High (public transport)
	The service and application market will grow	
Few services	○ ○ ○ ○ ○ ○ ●	Many services
	User security, integrity, and privacy will become more important	
Security a problem	○ ○ ○ ○ ○ ● ○	Security no problem
	Real or perceived health problems will become more important	
Radiation a problem	○ ○ ○ ○ ○ ● ○	Radiation no problem

Figure 2.1 (*continued*)

Environmental issues will become more important

Problem ○ ○ ○ ○ ● ○ No problem

Spectrum will become an increasingly scarce resource

Shortage of spectrum ○ ○ ○ ○ ○ ○ ● Abundance of spectrum

The wireless industry will grow

Slow growth ○ ○ ○ ○ ○ ○ ● Fast growth

The big NICs will continue their positive development

Slow growth ○ ○ ○ ○ ○ ● ○ Fast growth

Market concentration in the wireless industry will change

Few large actors ○ ○ ○ ○ ○ ○ ● Many actors

The fight for market dominance in the wireless industry will intensify

Operators dominant ○ ○ ○ ○ ○ ○ ● Operators lose

Infra. vendors dominant ○ ○ ○ ○ ○ ● ○ Infra. vendors lose

Term. vendors dominant ○ ○ ○ ○ ● ○ ○ Term. vendors lose

Terminal usage time and complexity management will become problems

Usage time a problem ○ ○ ○ ○ ○ ○ ● Usage time no problem

Complexity a problem ○ ○ ○ ○ ○ ○ ● Complexity no problem

3G will be implemented

Failure ○ ○ ○ ● ○ ○ ○ Success

Protecting IPR on content will become increasingly difficult

Effective protection ○ ○ ○ ○ ○ ○ ● Failure of protection

A Sunny Berlin Day in 2015

Sara finished her lunch with the people from the startup EmBase and left
the restaurant. Passing the entrance, she had a quick glance at the bill as
it popped up on her handheld and accepted it together with the suggested
rounded 8% tip her PDA agent had learned that she gave when having lunch
in Europe at restaurants rated between 20 and 25 by Zagat Surveys.

At last she had the rest of the day for herself after a morning full of
meetings. EmBase seemed to be on track with their new product and she was
going to recommend her investment firm back in Stockholm to finance another
round. EmBase is positioning itself for a new market using implanted sensors
to continuously monitor personal health status. In 2015 it is technically
feasible to swallow or inject nanosensors that will function in the human
body for months and transmit data about levels of its 50 most important
substances. EmBase is developing a PDA software package that will use all
this data to assemble a personalized health profile. The innovative part is
that EmBase will use the PDA to assemble more information by asking the
user directly and by automatic extraction of data on indicators already learned
by the PDA agent. This is information such as food intake, physical exercise,
sleeping habits, and subjective well-being. Over time the EmBase software
will be able to develop a very precise personal profile of the most important
body functions. It is then possible to understand the connections between,
for example, feeling dizzy and specific physiological indicators.

A warm spring breeze filled a sunny Berlin as Sara went for a sightseeing
walk. She was interested in the history of the city and decided to learn a little
about the divided Berlin during the Cold War. She put her GPS-enabled
projection goggles on and her agent projected her personal start screen in the
semitransparent eye field. She spent about a minute looking at her headline
news, the Berlin weather for the next six hours, a few local offers and events
her PDA had accepted, plus the five restaurants recommended by her agent
for the scheduled dinner tonight with Michael. She found one restaurant
interesting and had a quick look at recent pics from the restaurant webcam
to get a flavor of the atmosphere. Sara liked it and used the voice control to
ask her agent to make the suggested nonsmoking table reservation for two
people.

Ignoring her mailbox, she changed to the Berlin city guide and paid the
€5 fee for the 3D guide, which included the cost of the heavy data downloads. The guide menu for screen-goggle users popped up and she chose the
narrative guide for the Cold War. The screen changed and suddenly she was

transported to Berlin in the 1960s. Interlaced over the Berlin street where she was standing, she saw projections of how the buildings used to look; in the streets she saw people and cars from the 1960s reconstructed from various photo archives. She turned around and was almost shocked by the projection of the wall standing just 10 meters in front of her. Sara took off her goggles for a moment and looked at a peaceful street with no visible sign of its history. She put the goggles back on and a voice-over started telling the story of the place; it suggested she should walk slowly one block up. There another 3D scene started playing, reconstructed from films of the mass meeting with President Kennedy in 1963.

After telling the goggles to tone down the guide, she walked around for a while. Suddenly a sign was projected on the goggle screen over one of the women's fashion shops. It said they had a blouse in her size. She went in and her PDA agent transmitted a request for the item. The shop attendant received the specification at her counter and went for the blouse. The size was perfect but Sara didn't really like the style. "What can you expect?" she thought. "PDAs can get the basics right but you just can't teach an agent style." Sara liked the new system of sizing clothes. You go into a kiosk in a clothing store, undress and turn around while several lasers read your measurements and store your 3D body profile with perfect accuracy.

Sara started to feel hungry and wanted a salad. She hadn't been in Berlin for years and turned to the city guide again. She had put the goggles away and looked at her PDA instead. It projected a map of the nearest blocks and when she clicked on "salad type" places it marked the ones matching her profile, serving organically grown food that was genetically enhanced for higher levels of bioflavonoids and vitamins. The closest was just around the corner. Using voice commands for your PDA in places like restaurants is considered rude, so after finishing the salad, Sara used the pen to open her mailbox on her PDA screen. She had 12 live mails and the first words of the transcripts were displayed on the screen. Most were voicemails but one was a videomail from the VP of Werdieer in London. She must have passed somewhere with cheap bandwidth during the last hours as the videoclip was already downloaded in her PDA. She put in her earphone and opened it. He seemed to be sitting in the back of a taxi. He said hello then talked about the status of their alliance negotiations for about 10 minutes and enclosed some slides with diagrams. One of the other mails was a movie demo for a marketing film for one of her other projects. The file was 5 gigabytes, so the PDA had not downloaded it. The indicated fee for download at this spot was €12 and it would take over 7 minutes. "How irritating," Sara thought, "you never have enough bandwidth and disk space." She was thinking about downloading a compressed version for the PDA screen but decided to wait. She wanted to see the film with full quality.

Sara left and went back to the hotel. Once in the hotel room, she logged in with her PDA, which did the handshaking with the room and its fixed connection. Her personal desktop start screen popped up on the 160×90 centimeter screen on the wall, showing the Berlin weather and a reminder about her dinner engagement. Now she could download the marketing video file from her mail server in Stockholm on the fixed net. The high-def film was crystal clear and she zapped between the different angles available to her. Sara made a few comments about the film and replied. Then she looked through the rest of the inbox, replied to some messages, made a few voice calls, and video-called the Werdieer VP. He was on a flight somewhere over the Atlantic with his PDA on the air-seat table in front of him. After talking with him, she minimized her mailbox and started CNN and a local Berlin news channel in two separate windows on the screen, with the sound coming from the Berlin channel. She changed for dinner and only brought the voice phone unit as she left the room.

When she returned to her hotel room, her PDA agent had reduced the temperature to 18 Celsius, matching her personal preference profile. Sara felt like watching a film. Hardly any hotels had the expensive pay-films any longer as most guests skipped them; they just went on the net and downloaded pirated films. She went to FilmWarez, one of her favorite sites and checked what was available now. Sara chose a film and started watching. It lacked the quality of the official version but who wants it at Hollywood's rip-off prices? The film was really bad, so she turned it off after about 20 minutes. Instead she went to a website for open-source films. Here you could find shorter, mostly animated films by amateurs. After watching a short novel, she turned off the wall screen and voice-called her friend Lisa, who was living in San Francisco. They talked for a while before Sara extinguished the light and went to sleep.

The Wireless Scene in 2015

Rapidly Growing Industry

The economic downturn in the early years of the century slowed industry growth for a few years. However, the rapid technological development within the communication and information technology industries continued and essentially all markets and industry segments experienced more or less continuous growth. Markets increased in size and scope. The operator industry consolidated at the same time as new types of competitors emerged. MVNOs (mobile virtual network operators) flourished, as well as local and regional players. Alliances of different kinds were formed. The sectors showing the highest growth are the service and application industries. The demand for advanced and demanding services (bit rate, QoS, etc.) is very large. People view their terminal as an extension of themselves and a multitude of services are offered to all segments. There seems to be no end to humans' need to communicate.

Industry Fragmentation—Market Leaders Losing Hegemony

Wireless services and applications became a huge success and in a rapidly transforming industry the old market leaders lost their dominant positions. The old telco world with closed, vertically integrated solutions gave way to layered open architectures based on Internet Protocol (IP) and the datacom industry won the market battle. However, in a maturing industry, profit margins were squeezed and the datacom market winners could never really reap high profits or leverage their market power. Open standards such as the internet and release of new unlicensed spectrum undermined the established market leaders. The incumbent players consolidated but in a maturing IT/tech industry, profits were eroding away as their markets became low-margin commodities. Independent consumers, open-source software and do-it-yourself wireless access further undermined corporate hegemony.

The market leaders did not vanish but the rapid technological development was as ruthless in turning profitable products into low-margin commodities as it had been in creating the markets. Mergers and consolidation reduced the number of players but their market power was nevertheless greatly reduced. Industry fragmentation and vertical disintegration accelerated when companies became more and more specialized. When performance of any given technological function was good enough, design and manufacturing knowledge around this function was no longer a critical asset and modularization set in. Consequently, this part of the market split into several new niche markets.

Now in 2015, users have had two decades of using the web, the PC, and the cellphone. Hence familiarity with these technologies in the general population is very high. Advanced consumers have been very active in trying out new technologies to bypass operators and other large corporations. And when successful solutions were good enough, the mass market followed. Deploying networks in unlicensed bands became very popular and undermined the dominance of traditional operators. WLAN access points evolved into do-it-yourself (ad hoc) wireless networks.

Portals gained significant market share but failed to own their customers in a closed system. The IPR lobby tried to enforce their rights but respect for copyright was eroding in the general population. Feeling this value shift, governments were more and more reluctant to help the IPR industry.

Debt-Burdened Operators Losing Market Dominance

In the telco financial crash in 2004 and 2005, many operators were brought close to bankruptcy. Some went bankrupt and the survivors picked up their network assets, but the industry as a whole did not get rid of its heavy debt burden. By keeping tariffs high, operators drew regulator attention to the industry, who in turn mandated that MVNOs would be given favorable access terms for using the networks. MVNOs took a significant and profitable part of the market by teaming up with successful brand owners possessing large customer bases and by keeping tariffs well below the levels of traditional operators. MVNOs had lower technical competence than the operators but cheap voice plus no-extras IP data access proved a good enough offer for many users.

When wireless data started, traditional operators first tried to offer closed telco-style services and developed in-house wireless portals. Seamless roaming (as in the voice GSM world) was very hard to accomplish with wireless data over a number of different underlying networks. Due to their slowness and poor execution, the operator offer of "value-added services" was regarded as bad, inferior, and expensive. This opened the market for independent wireless portals and services. The winners had attractive offers and gained a large user base. The operators were slowly reduced to bit-pipe providers. Customers trusted wireless portals, consumer brands, and MVNOs more than the operators.

But the major blow to operator dominance was the rise of unlicensed spectrum and WLANs. By keeping high prices for wireless data, operators opened a market for WISPs based on WLANs. It started as a large number of small WLANs deployed by IT departments in companies, schools, and hospitals. Then came WISPs (wireless internet service providers) for hot-spots and consumers putting up WLANs at home. In a bottom-up process, all these small local networks slowly merged by joining clearinghouses for AAA (authentication, authorization and accounting) and billing. Suddenly operators faced a large virtual competitor offering spotty internet access but at very low tariffs. Many operators refused to include WLAN offers in their services, therefore many customers left to join MVNOs and others who offered a full package of wireless services.

Telco Equipment Vendors Lose to Datacom Attackers

Traditional telco equipment vendors failed in responding to a rapidly changing marketplace. Telco vendors were adapted to a business model built on selling extremely expensive systems to a few very demanding operators and they were dragged down together with their traditional customers. The mobile operators themselves had aggressively asked for application program interfaces (APIs) and wanted to get out of lock-ins requiring them to buy large and complex proprietary systems. When vendors started delivering system modules, very high profit margins on for example base stations became visible and price competition plus new entrants in the market segment overturned this once lucrative market. Low-end market attackers from the datacom industry and from NICs such as China began eating away market share with very low prices and a business model built on low-cost, high-volume production. Advanced functionality such as sophisticated spectrum management was not that critical when large chunks of new spectrum were released. Traditional vendors were blind to this threat from disruptive innovation and believed they were safe by selling high-end proprietary equipment to their old-boys network. The hardware market for base stations, etc., underwent significant structural changes during 2005–2010. Assembly became a low-margin market for contract manufacturing firms. Design was gradually moving into silicon chips and a growing slice of profits went to the processor makers. All this meant lowering of barriers to entry and turning wireless hardware into a commodity market with market leadership based on volume. The base station market itself was undermined by cheap wireless access points, originally WLAN access points but evolving into multipurpose install-it-yourself wireless access points.

With modules and open APIs from the datacom world, the problems in putting networks together were almost removed. This opened the market for vendor-independent systems integration (SI). Telco vendors correctly identified SI as a growing market segment and had some success with their traditional demanding customers. However, the SI arms of the vendors were initially too vendor biased and lost credibility compared with independent SIs. Also, the independent SIs targeted growing market segments such as MVNOs, WISPs, corporate WLANs, and clearinghouses. The only high-profit market dominated by traditional telco vendors is now the core systems market (the complex software controlling the intelligence in the traditional mobile networks). However, this is now mostly a shrinking legacy market, with customers from locked-in traditional operators. All emerging operators with simpler and more cost-effective business models had to do without the complex core system software. They had to rely on less functional but radically cheaper (or free) software, based on IP, SIP (Session Initiation Protocol), and other open APIs. Some of these software modules were free, or included when buying base stations or other hardware. The disruptive innovation undermining this market was the do-it-yourself plug-and-play wireless access box.

Terminal Vendors Attacked from NICs and Datacom Industry Vendors

Telco terminal vendors lost market power when the commoditization of the market occurred more rapidly than expected, dominated by open IP access. The critical telco knowledge embedded in the radio and codec (coding and decoding) software was gradually commoditized by attackers from Japan, the NICs, and the datacom industry. As the WLAN movement kept growing, WLAN access points expanded into wireless access of various forms for do-it-yourself use. In this process, an open-source movement started producing free software for the codec and radio interface.

The handheld PDAs saw a rapidly growing market, boosting the ability for knowledge workers to access their Microsoft Office documents when outside the office. Laptops became more of a stationary computer and many knowledge workers used smaller handhelds when on the move. As corporate IT departments insisted on Microsoft Office compliance, traditional telco terminal vendors were excluded from this market.

The handhelds together with laptops, machine-to-machine communications, and game terminals created a large market for wireless internet access. Initially, these gadgets had traditional mobile voice units included, licensed from the telco vendors. But eventually it became possible to offer acceptable quality with VoIP. This did not totally replace traditional mobile voice, but for most noncritical terminals, VoIP was good enough and reduced licensing revenues and market dominance for traditional terminal vendors. Market power shifted to handheld makers together with processor and component manufacturers.

Active Users Driving Development and Undermining Copyright

In the first decade of this century, users were very active and drove this development towards an open IP world with skyrocketing traffic and an abundance of applications and services. They preferred freedom and choice over convenience and didn't accept being locked into corporate bundles.

For people born around 1990, using the cellphone and the net was as natural as breathing. This generation, the Moklofs (mobile kids with lots of friends), is now the new young user segment and they know inside out how to use the technology to their advantage. As kids grew up without having a lot of money, their values became very much the values of the underground culture, where you don't pay for anything and you find smart ways around big company hurdles. Open-source software is an integral part of this culture. The attitudes towards copyright and IPR (intellectual property rights) in this generation became very negative. The IPR industry had lost credibility by trying to push its interests in a maximalistic way. Instead of embracing the net as a new super efficient distribution channel where negligible costs make it possible to profitably sell content at near-zero cost, the IPR industry went for a high-price strategy, further undermining consumer acceptance of paying for content. The IPR industry fought a retreating battle by trying to ban all new technologies such as DVD burners and in 2007 it managed to get government support in forcing ISPs, WISPs, mobile operators, PC makers, and terminal makers to install copy-blocking functionality. After that, the IPR industry dramatically raised prices. The public backlash was massive and fixes to undo the copy blocks spread like wildfire. This was the moment the underground culture became mainstream. Not that crime was accepted, but rather that downloading copyrighted material for personal use was considered to be a part of freedom of information and expression. Governments, feeling the shift in public opinion, were reluctant to chase voters on this matter and by 2015 the IPR industry still hasn't formed a viable strategy for the future.

In a society of material abundance, values gradually shifted towards deriving esteem from giving away what you could dispose of without being worse off yourself. For example, allowing others to temporarily hook up on your wireless access point or giving away (i.e., transferring to others) movie files from your hard disk.

A Mobile Lifestyle with Increasing Travel

The trend of increased travel continues without pause. The urbanization trend continues to expand globally. More and more people and companies move into cities and their surrounding areas. The result is larger cities but also the birth and growth of regions surrounding them. In the wake of this development, by 2015 most transportation systems are in the process of being upgraded. Local governments and companies invest in new and bigger public transport systems and road infrastructure to pave the way for a highly mobile workforce. The dominant lifestyle is nomadic; people are willing to move to where the jobs are and they accept a lot of travel. Time is a valuable asset and everything that saves time is welcomed. Efficiency is a key word.

Home working has increased among knowledge workers and middle and upper managers that can do some of their work independently. The result of this development is an increased number of people traveling longer distances to work. However, the average time spent commuting is not higher, mostly due to the development of faster transportation systems and better road infrastructure. As a result of the development of new and better public transportation systems and the fact that people spend all their time using just one mode of transportation, people are spending their traveling time working to a large extent.

For those commuting by car, numerous new services emerged but of a different kind than those in the public transportation systems. These services include travel information (smart roads and cars, intelligent traffic systems), conferencing while on the move, and synchronization of information with co-workers at the office. More and more people share cars to and from work and there are even a number of commercial agencies for this, using wireless networks for coordination of the car fleets. For car-poolers, the journey can be used more effectively, at least for the passengers.

An Explosion of Services and Applications

In the industrial countries as well as in the most successful NICs, cellular systems are complemented by a large number of other systems (e.g., ad hoc networks, WLAN access, satellites, high-altitude platforms). Most technical problems concerning seamless roaming, system integration, etc., have gradually been solved.

Appetite for wireless applications and services is very high, and once the geographical positioning infrastructure was in place, the number of location-aware applications and services grew rapidly. Wireless services are used by everyone in all segments. There is a large variety of usage patterns and services. A whole range of different terminals are used, ranging from simple voice and messaging only, to very advanced and powerful computer-like terminals with advanced input/output technology. Wearable communication devices are common and there are thin clients used in various context-dependent and intelligent networks. Platforms and protocol are standardized and open.

Professional users and Moklofs drove the market the first years. Gradually, Industrial users and then other consumer segments followed. Besides voice, the early killer application was messaging that soon evolved into full email functionality. Messaging and email soon converged, using unified mailboxes. A real hit was sending digital pictures from terminals with built-in cameras, and the wireless messaging scene soon developed features such as communities, group mails, chat rooms, flirting, dating services, etc. Entertainment proved to be a successful market with games and gambling as the leading services. Other successful applications in the early market were time-critical services such as weather forecasts, stock market trades, online banking, and directories.

When advanced enabling technologies such as geographical positioning, voice input/output, and secure payment systems became available, the application market took the next step. Adding location information gave a new dimension to all mobile applications. Communication and messaging were greatly enhanced by adding real-time positioning, e.g., for buddy lists, communities, role-playing games, flirting, and dating services. Location-based services such as maps with driving and walking directions to the nearest open restaurant or gas station were also an instant success, as were all other information and transaction services.

Super advanced terminals such as goggles with projection screen and video camera are available in 2015 and are almost on the mass market. The goggles interlace a full screen in the field of vision and are used by Industrial users, medical professionals, and early adopters. Popular consumer applications use very accurate positioning of the goggles and interlace images over the real-world view. When moving the head, the projected image follows in real time, allowing users to project how a tourist site looked 37 or 258 years ago, with a storyline about past events.

There has been an explosion in the number of companies developing services and applications as well as companies delivering services and owning content. But this industry can be described as rather fragmented. A few very large companies dominate, yet only for certain services, in certain locations, and in certain segments. There is plenty of room for small independent companies as well. Some of these small companies operate in very local markets, others focus on niche services and narrow customer segments. Various types of brokers, middlemen, and wireless portals exist. Even though many services don't rely on specific content, brand is an important aspect. In this service-rich world, brand is a good way to attract customers, and it is even more crucial to have properly branded content. The big content owners (CNN, AOL Time Warner, Disney, etc.) provide all kinds of service providers with content. Large entertainment, gaming, and gambling companies have also emerged on the wireless scene.

Spectrum—Abundant Release for Unlicensed Bands

During 2005–2010 governments released significant chunks of new spectrum from TV broadcasters and the military. Spectrum was allocated mostly for unlicensed use but also to traditional operators. To avoid the auction debacles, new and more refined models were developed based on revenue sharing during the productive time of the new networks. With much more available spectrum, traffic prices fell faster and the dominance of the incumbent operators was reduced.

Unlicensed spectrum usage was a huge success with skyrocketing traffic. The unlicensed bands drove rapid innovation of cheap install-it-yourself "black-box" access points that can double as multiband base stations. These black boxes can be used to offer access for a number of wireless bands and technologies. Some radio bands were illegal to use but often black boxes were sold in Europe and the US with the illegal parts just disabled, a problem hackers soon overcame. In less regulated countries they were sold openly and leaked back into Europe and the US. Eventually regulators bowed to the facts and eased regulations. After initial resistance, operators allowed them to be integrated into their own cellular structure by offering standardized interfaces and revenue-sharing schemes. They saw the opportunity to increase coverage without bearing the cost of building the infrastructure. Unlicensed spectrum release was a retreat battle of vested interests. Some countries released more spectrum and created a market for applications in these bands, thereby setting an example that other countries had to follow.

No Real Problems with Integrity, Privacy, and Security

Privacy and security are not regarded as a large problem and well-informed users now know what to avoid. New software offers fairly efficient protection for viruses and other traps. The information explosion and media saturation drew development toward a more transparent society—a society in which people can find out everything about each other if they bother to look for it. In a more transparent society, privacy is viewed as less important than all the benefits you can have from these new services. On the few occasions when security is really important, technologies such as biometric identification and encryption have been developed to provide the desired privacy.

Fast Development in China and Other NICs

Developments in China and the other NICs have been very positive since around 2005. Gradually these markets have been deregulated, political unrest has been avoided, and finally the financial strain on their economies has diminished. These countries gradually decreased the gap (technology, usage, etc.) between themselves and the industrial world, and they are now important players in terms of markets and technological development. A few of the low-end players in the global telecom market, both operators and vendors, are based in these countries. Large infrastructure investments have taken place in these big NICs. Even some developing countries have invested quite heavily in cellular systems, and much of the world's population now has easy access to at least a telephone. Several of these countries skipped the fixed networks and invested directly in mobile systems, since this gave considerable cost advantages.

Batteries and Complexity Management No Showstoppers

The lifetime of batteries for mobile terminals has increased dramatically since the turn of the century. Batteries are now used on a large scale for an enormous number of services and applications, which has led to very large production volumes and price drops for these new energy sources. For the same size and price, capacity has increased by at least five times. Systems for efficient recycling of the batteries have also been introduced. The business world has especially welcomed this change. Employees are now able to move out of the office and still use the same services and applications as on their desktop computers. This increased usage of portable computers increases the need for wireless access to the network, boosting the development of various wireless technologies. The development of battery technology has also boosted services for other user segments, making true mobility a possibility for most types of service, even those consuming large amounts of power. For most users, battery lifetime is no problem.

Research is now mostly focused on decreasing the inherent danger of using ultra high capacity batteries of small size (over 10 ampere-hours for AA size). For the time being, these are available only for some professional and military users with special needs.

The past decade has seen the implementation of many different wireless systems and technologies. In the Western world, as well as in Japan and some of the big NICs, cellular systems of different generations (2.5G, 3G, etc.) coexist with other types of system (WLANs, PANs, BANs, broadcasting, satellites, etc.); some of these are self-configuring whereas others require extensive planning. A major difficulty has been to manage the complexity of this inhomogeneous wireless world. In general, these difficulties have been solved. Integration of different wireless technologies and roaming are possible in almost all dimensions. The operators still control the bit-pipes of traditional networks and have full control over their performance through complex control systems that enable data to be collected and processed centrally. However, the big problem has been the integration of old traditional networks and the new emerging networks such as those operated by WISPs, corporate WLANs, ad hoc networks, and self-configurable networks. This integration is very dependent on IP and other open standards, enabling simple wireless black boxes to be directly plugged into the wireless infrastructure. Issues concerning transmission security and roaming have also been solved. Very secure services are provided by VPNs (virtual private networks).

Wireless Technology in 2015

A World with Many Different Wireless Systems

A range of systems provide wireless communication. In this heterogeneous world, wireless access is provided by various public and private networks, shared between operators and seamlessly integrated for the users. This means that the wireless world is characterized by many air interfaces ranging from a few kilobits per second to 100 Mbps and more. The radio access infrastructure is very dense in urban areas and much of the traffic is asymmetric—users download much more information than they send. Resource management and control of the systems are often distributed, especially for systems on lower hierarchical levels. It is essentially only the large cellular systems that remain centrally managed.

In the public cellular networks, 3G is living in parallel with the old 2G and 2.5G systems but 3G has taken over a significant part of the traffic volumes. In many regions the 3G systems have evolved into more advanced systems, giving substantially higher capacity (3.5G) and the world is now on the verge of taking the next step. These systems, based on macrocell structures, provide good coverage with data rates above 256 kbps in rural areas. In more densely populated areas they are complemented by micro- and picocell systems enabling data rates several times higher. Radically new air interfaces are under development in the lab, with possible commercialization around 2020. Coverage is also provided by various broadcast and multicast systems, with very high capacity in the downlink. High-altitude platforms (HAPs) and satellites provide communication for vehicles moving at high speeds (cars, high-speed trains, etc.) and users in very remote areas such as ships. In many rural areas, where it is too costly to provide access with fiber, fixed wireless access (FWA) is common for the last two miles. For outdoor cellular systems, traditional network planning techniques still dominate. Indoor and also some urban systems (picocells, etc.) are deployed ad hoc by the users themselves or by various service providers. Roaming between all these systems is possible and not really considered a problem anymore. For the user, the wireless world is transparent and seamless.

The cellular systems are complemented by very flexible systems providing coverage in smaller areas, both indoors and outdoors, with high demands on capacity. A typical WLAN hotspot has a shared capacity of 500 Mbps in the covered spot. The WLAN systems are deployed by WISPs or most commonly by local WLAN owners who sell capacity to operators, service providers, etc. There is also an abundance of very short-range systems such as personal area networks (PANs) and body area networks (BANs). These use Bluetooth or similar standards and make possible any type of data transfer in close proximity to the user. Multihopping, using the terminals as relaying nodes, is quite common. WLAN and Bluetooth-like technology is embedded in all personal computers, PDAs, digital cameras, etc.

The cost of base stations, access points, etc., has decreased dramatically during the past decade. Combined with user deployment, multihopping techniques, distributed resource management systems, etc., the cost for designing, building, and maintaining the infrastructure has decreased substantially, even for cellular systems.

The terminals range from very low end to very high end, from cheap to expensive. The most advanced can handle very high bandwidth and have advanced user interfaces based on voice recognition, etc. Terminals may be classified according to the number of services they can manage and the quality of service they provide. Battery lifetime is at least a week, even for heavy use.

The user interface ranges from simple to complicated depending on the user's requirements and knowledge. Voice and other means of controlling the device are fully developed. Some terminals are dedicated to a specific service, e.g., voice and messaging only. Others are multiservice, support different air interface standards, and can handle several frequency bands. Terminals are everywhere—in clothes, in jewelry, in goggles with a built-in screen but also in all conventional terminals such as PDAs, voice-only phones, and laptops.

An Abundance of Services with Various QoS

Content is personalized according to user demands and location. Prime services are MMS, videoconferencing, media broadcasting, in-vehicle services, banking transactions, corporate LAN access, time-critical news, M2M, etc. The most important quality of service (QoS) factors are latency (in milliseconds), geographical availability (coverage), temporal availability (uninterrupted uptime) and service while the terminal is moving at high speed. QoS is typically provided as a function of the price. High-priced services can provide end-to-end latency for data well below 100 ms and voice latency below 170 ms, full urban coverage including indoors, and availability 99.7% of the time (26 hours of downtime per year). The higher the price the user is willing to pay, the higher the QoS provided. This means that QoS is not guaranteed for free or low-cost services. Due to the heterogeneous nature of the wireless systems and the wide array of services, there are many communication clearinghouses providing users with a package of service solutions.

Standardization Has Increased

Standardization has increased dramatically in all areas, e.g., software, terminal components, protocols, and air interfaces. This has been very important in order to combine all the different types of networks and terminals. De facto standards dominate. All traffic is IP based in wireless networks and fixed networks. The software in most terminals is based on open-source and standardized operating systems, etc. New services are easy to develop and introduce since they are developed on open application platforms. Standardization gives us automatic handshaking and plug-and-play everywhere, hiding all complexity for the user.

3

Slow Motion

Key phrases: slow technological and industrial development, global recession, radiation a health problem, environmental awareness, hacking and security still a problem, industry consolidation, no service explosion, big NICs catching up

The wireless world has developed at a slow pace since the turn of the century. The global economic recession during the first decade plus real and perceived health problems due to radiation from wireless devices and equipment deeply affected the wireless industry. Even though the demand for mobile services has increased, the service explosion that many actors and companies envisaged never materialized. The communication industry in general and the wireless industry in particular have gone through substantial change. Consolidation has increased and the number of companies in each market has been reduced. Technological development has slowed down and profit margins have decreased substantially. The industry has matured and can no longer be considered a hi-tech industry. The big NICs, for example China, India, and Russia, are catching up faster than industry had expected, since their recovery started earlier than in the industrialized world. Currently they are not only important as the biggest markets for wireless products and systems, but also because some of the most important global companies are based in these countries.

Figure 3.1 shows the 14 trends defining the scenario space and how we have set their values for this scenario. The scenario begins with a personal story of life in 2015. This story is followed by the main scenario, describing important aspects of the wireless world in 2015. The chapter ends with a brief description of the technical systems and services used.

Figure 3.1 Slow Motion: defining dimensions. All variables are described in Chapter 6; all user segments are described in Appendix A

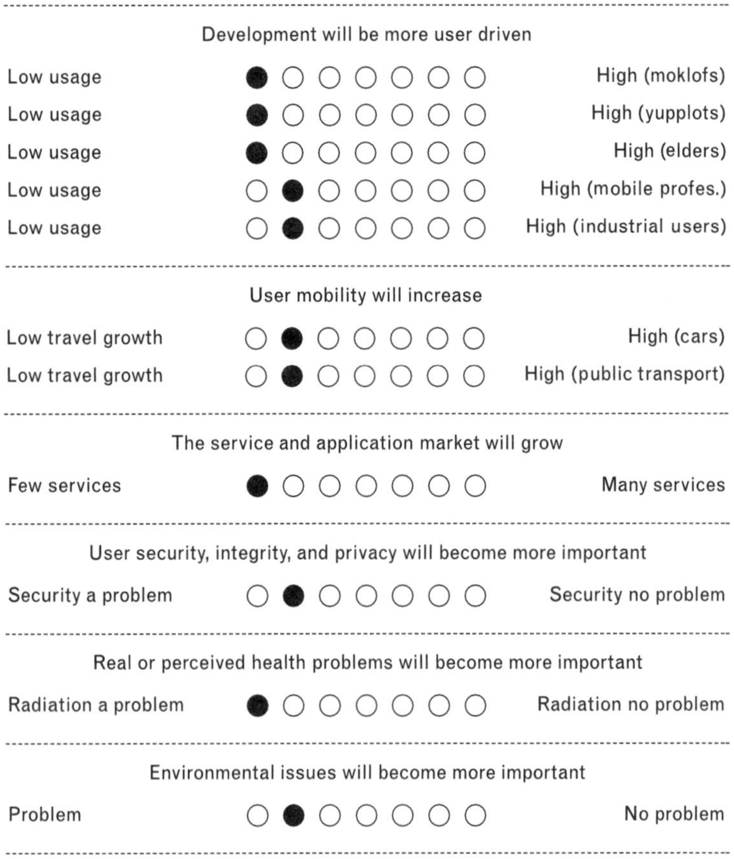

Figure 3.1 (*continued*)

Spectrum will become an increasingly scarce resource

Shortage of spectrum   ○ ○ ○ ● ○ ○ ○   Abundance of spectrum

The wireless industry will grow

Slow growth   ● ○ ○ ○ ○ ○ ○   Fast growth

The big NICs will continue their positive development

Slow growth   ○ ○ ○ ○ ○ ○ ●   Fast growth

Market concentration in the wireless industry will change

Few large actors   ○ ● ○ ○ ○ ○ ○   Many actors

The fight for market dominance in the wireless industry will intensify

Operators dominant   ○ ○ ● ○ ○ ○ ○   Operators lose
Infra. vendors dominant ○ ● ○ ○ ○ ○ ○   Infra. vendors lose
Term. vendors dominant ○ ● ○ ○ ○ ○ ○   Term. vendors lose

Terminal usage time and complexity management will become problems

Usage time a problem   ● ○ ○ ○ ○ ○ ○   Usage time no problem
Complexity a problem   ● ○ ○ ○ ○ ○ ○   Complexity no problem

3G will be implemented

Failure   ● ○ ○ ○ ○ ○ ○   Success

Protecting IPR on content will become increasingly difficult

Effective protection   ○ ○ ○ ⊞ ○ ○ ○   Failure of protection

= Not applicable or considered in scenario

Ordinary Life in Stockholm and Business Life in Shanghai

A Day in the Life of an Ordinary Swede

In Stockholm, Anders Nord is on his way home from the construction site where he is working. He is a carpenter and has worked almost two years in the new community being built about 20 kilometers northwest of the city. Even though he often complains about his work, he is quite happy that he chose the construction industry after graduating from high school. At least it meant that he has always had a job, even during the tough years in the middle of the previous decade. Many of his friends were laid off following the bursting of the great telecom bubble, which triggered a global recession rapidly spreading to many other industries.

Anders listens to the radio while keeping an eye on the traffic management system, an expensive extra in his Volkswagen. The information is updated once every 15 minutes. It seems there is no major congestion or roadworks on his way to his apartment in a suburb south of the city. Since he has driven this route more times than he cares to remember, he doesn't use the GPS module of the traffic computer. Actually, he seldom uses it even if he is in a new city, since the service is quite expensive.

Just as he turns into the parking lot, he receives a phone call. He fiddles around a little bit with the earpiece but manages to take the call before it rolls over to the voicemail. It's a friend and after a short conversation they decide to meet later in the evening for some food and a movie. Even though Anders recently bought quite a nice home entertainment system, he enjoys watching the latest releases in a movie theater.

After a quick shower, he checks his messages on his mobile terminal, a device with basic functionality such as voice, messaging, and simple surfing. The model he uses is not so different from the first mobile phone he bought around 2002. It radiates a lot less and it is easier to use, but on the whole, terminal technology has developed slowly. He also remembers to check the latest traffic information from the commuter rail company. No major delays today it seems. Normally he uses the fixed computer to surf for information and entertainment, but since this is only a single and simple task, he uses the mobile terminal. Checking his watch, Anders realizes that he has another 15 minutes before he needs to leave. He switches on the computer and downloads parts of the afternoon edition of the newspaper he subscribes to. He chooses the sports and entertainment sections in SMTF (Simple Mobile Terminal Format) and then uses the IR (infrared) port to transmit the content to the mobile terminal. This saves him money and time compared to downloading it directly to his terminal or reading the news online.

When entering the railway station, he writes a short code on his terminal and the green light on the gate comes on. He pushes through and enters the platform. The terminal informs him that this is the last day of his prepaid period and asks if he wants to renew the train pass. Anders sighs and replies yes, knowing his phone bill once again will be much higher than expected. On the train to Stockholm City, he reads the sports news and as usual he becomes depressed by the poor performance of his favorite soccer team. He also checks the movie section and finds a couple of films he would like to see.

Anders meets up with his friend at their regular sports bar and orders the usual burger and fries. Knowing they are in a WLAN hotspot, they use his friend's PCIA (personal communication and information assistant) to download a few short videoclips from the movies they have picked out. The PCIA is a standard model intended for nonprofessional users. It is not the most advanced product on the market but it serves its purpose reasonably well, even though it is slow and a bit cumbersome to use. After some discussion they decide on one of the movies and order the tickets. The PCIA answers with a reservation number and informs Anders' friend that the tickets will be charged to his phone bill. Arriving at the theater, they give the number and enter.

Going home on the train, Anders relaxes and listens to some music he downloaded yesterday and stored in his terminal. This can still be done for free, if you know where to look on the net. Halfway home, a voice interrupts informing him that the battery is running low. He turns off the music, saying to himself, "Wireless communication technology still has a long way to go before it reaches perfection."

A Business Day of a Mobile Professional in 2015

Mr. Lee looks at his watch. It's almost two o'clock and it's time to go to the Blue Dragon conference room at the headquarters of Great Wall Telecom, the Shanghai-based telecom operator and service provider. Great Wall is a global company that has grown very rapidly during the past 10 years and is currently the world's fourth largest operator in telecoms and datacoms. Mr. Lee is chief technical officer (CTO) at the European headquarters in London; he flew into Shanghai the day before. Since becoming CTO about a year ago, Mr. Lee has begun to refocus the infrastructure part of the company on enhancing the 2.5G systems to 3G. Great Wall owns and operates several 2.5G cellular networks (and even some old 2G networks) in Europe as well as a few 3G systems built between 2004 and 2006.

He turns off the stationary computer at the desk he has borrowed during the day. He grabs the briefcase in which he carries his PCIA, a state-of-the-art wireless device for Mobile Professionals such as himself. The terminal is connected to a wireless earpiece attached to his glasses. The earpiece is an important part of his body area network (BAN) and it also reduces the level of electromagnetic radiation close to his head. While walking down the corridor, a soft voice informs him that he has received an email from his wife, but he decides to read or perhaps listen to it later.

Arriving at the conference room, he greets his colleagues from Great Wall as well as three representatives from the world's largest telecom equipment vendor, Constellation, formed in 2008 by a merger between two of the remaining European infrastructure vendors. Today's meeting is to discuss the results of final testing and the rollout plans for the 3G systems to be deployed in Europe. The new networks will increase capacity and quality of service, as well as reducing the electromagnetic radiation levels. This is a very important contract that Constellation received in competition with its main competitor, General Telecommunications.

Mr. Lee sits down and turns his PCIA to meeting mode, which means that only emergency calls or messages related to this particular meeting are let through. He also makes sure that it connects to the videoconferencing system in the room. On the large wall screen, as well as on his desktop display, an image of himself comes up together with a text message saying that he has now entered the meeting. A quick glance on the screen notifies him that the Great Wall team in California is also connected. Mr. Lee looks at Mrs. Smith from Constellation and asks about Mrs. Eloranta, the chief engineer responsible for the equipment that Great Wall has ordered. Unfortunately, Mrs. Eloranta's flight to Shanghai was delayed due to bad weather, and so she will participate from a hotel room in Bangkok. Soon enough, the videoconferencing system informs them that Mrs. Eloranta has joined the meeting. Mr. Lee lets out a sigh of relief and irritation. She has an important role in this meeting, but now they have to make do with the poor quality of her mobile videoconferencing system. Mrs. Eloranta excuses herself, saying that at least she can participate from the privacy of her hotel room using her PCIA. Luckily enough, Mr. Lee and Mrs. Eloranta are quite well acquainted. It's not the first time they have done business together.

The meeting proceeds rather well, even though one group participates from the videoconferencing room in Shanghai, one group from a similar room in California, and one person from a regular hotel room in Bangkok.

While discussing technical issues, the electronic whiteboard and the slide display on the computer screen are very useful. It is possible to show slides and other electronic documents as well as drawings done by hand. This unfortunately poses a problem for Mrs. Eloranta. Her PCIA is a state-of-the-art device and she can see the other participants on her laptop to which her mobile is connected. But the images of the other participants are small and their quality is poor. She can see the slides being presented and she can also present slides of her own. She cannot, however, see what is drawn on the electronic whiteboard. The delays are annoyingly long and she becomes increasingly irritated by the poor quality of the sound and video streams coming from California. "Thank God for 3G," she says to herself.

After the meeting, Mr. Lee takes a taxi back to the hotel. During the ride he checks and answers a few emails that he received during the meeting. He has also received a few short messages, one of which is a photograph of his youngest son playing in the pool in the backyard of their new house. Mr. Lee smiles, and as the taxi stops at a traffic light, he takes a photo of the Oriental Pearl, Shanghai's television tower. It doesn't come out perfectly but it's good enough. He jots down a few lines telling his wife and sons that he loves them and sends the electronic postcard. Back home in London, Mrs. Lee drinks her morning coffee and smiles at the electronic postcard that she reads from her standard terminal, a simplified version of her husband's PCIA. It's the third time she sees the Oriental Pearl, and this photo is even worse than the previous two he has sent her.

After about 60 minutes in the rush-hour traffic, the taxi pulls up outside the hotel. Mr. Lee uses his PCIA to pay the driver and walks into the hotel lobby. He finds a table in a quiet corner of the bar and orders a beer. He logs into Great Wall's intranet and checks the automatic calendar function. His assistant has booked a meeting for him tomorrow morning. He swears silently but accepts the meeting. For each long trip he makes, it becomes more and more difficult to adjust to the change of time zone. Finally, he checks the latest online flight information. It seems his late-evening flight back to London will be on time.

The Wireless Scene in 2015

Economic Recession and 3G Fiasco

The global economic downturn that started in 2001 turned into a large-scale economic recession, even a crisis in many regions and industries. The telecom, computer, and media industries were severely affected. The burden of gigantic debts from the 3G auctions and continued price pressure on wireless tariffs put enormous pressure on the mobile operators in the early 2000s. When adoption of new wireless services proved to be slower than expected, the financial actors refused to finance further investments. It started to turn really bad when one of the largest European operators, after a prolonged and desperate struggle, went bankrupt due to extremely large debt and negative cash flow. This spread very quickly in a domino manner to other operators and eventually to equipment and terminal vendors as well as to service developers and providers. Several large telecom actors disappeared and those that survived made massive cuts and saw drastically reduced margins. The media called this the bursting of the great telecom bubble. The loans to the telecom sector, especially to operators, were staggering. According to some calculations, they amounted to as much as 5% of the world's total GDP. The financial sector was severely affected when a large part of these loans turned bad. Several European banks went down, resulting in governmental rescue packages and a major restructuring of the whole financial industry.

When the assets of all the operators that went bankrupt between 2004 and 2007 were sold, the new owners renegotiated the old 3G commitments with European governments. Some networks were canceled and many half-built networks were merged, resulting in only one or two physical networks per country. In many parts of rural Europe there is still no 3G coverage at all. After the failure in Europe, many operators in other parts of the world hesitated to install 3G. In the US, the government never allocated any 3G spectrum at all, hence North America is still lacking 3G. In 2015, 3G is only a stand-alone service in Japan and a few smaller countries. In Europe and the rest of the world, the spotty 3G networks are subsumed under service bundles from the operators. Some 3G networks were closed altogether and the spectrum was reallocated to other wireless technologies. Since the existing large operators in the industrial world are still burdened by debt, they are now focused on updating and improving their existing systems.

Health Problems from Radiation

The long-term studies of how radiation affects human tissue that were presented around 2005 still have a negative impact on the wireless industry. The results were clear and most experts finally agreed that mobile phones and other wireless devices, when heavily used (more than one hour a day), would injure the brain due to radiation from the transmitter. Some studies, though somewhat inconclusive, also indicated a risk for genetic effects of the radiation. Other studies showed similar effects from base stations, especially on people working or living near the transmission source. In the beginning, the telco industry fought back by arguing that the results were inconclusive and referred to other studies that had reached less clear results. They unsuccessfully tried to convince the general public and authorities through joint and massive PR efforts and heavy lobbying. When consumer organizations in the US finally won a number of large lawsuits, things started to change. Facing staggering costs for medical care and dramatically reduced sales, the telco industry reacted in a proactive way and managed to avoid total disaster by suggesting very strict regulation of radiation levels from new wireless devices; it also took measures to drastically reduce radiation levels from new and existing base stations. The term "RF pollution" quickly became part of everyday language and some communities even began marketing themselves as "RF-free zones."

Even though manufacturers of devices and base stations questioned the research results in the beginning, they quickly started to redesign their wireless products in an effort to keep their worried customers. Since it turned out to be quite impossible to convince the general public that the products were safe, the use of wireless devices dropped dramatically. At the same time, mostly through connecting one fear to another, people also stopped using less dangerous wireless devices such as WLAN and Bluetooth, just because they also radiate and thus might be dangerous. Usage is still substantially affected, even though most of the problems have been solved through redesign of devices and other equipment as well as due to strict regulation of usage and radiation levels. It seems people fear what they cannot see, so they avoid using wireless devices if at all possible.

It took almost 10 years before recovery began, and by 2015 the communication industry has only just started to get back on track.

Security a Problem Still Waiting to Be Solved

The problem of hacking and virus creation is still significant. Despite great efforts from industry and governmental agencies, it seems a never-ending battle in which hackers and criminals are always at least one step ahead. Even though we have seen significant advances in encryption techniques during the past few years, most codes are quite easily broken with powerful quantum computers and new mathematical techniques. Viruses are easily spread in the various wireless networks, especially through multimedia messaging (MMS) services and when downloading software, and the average consumer experiences frequent software crashes and other annoying problems. The problems increased when various data services were introduced in the updated 2G cellular systems and were further accentuated with the introduction of 3G. Many people feel that they cannot trust electronic transactions and are seldom willing to e-shop. Nor do they feel secure when contacting the government and other bodies with sensitive information. Wireless services are especially sensitive, since anything you send by radio, you send to everyone.

The Mobile Lifestyle Loses Ground

In the industrialized Western world and Japan these problems meant that the mobile lifestyle in the private sphere came to a halt during the first decade of the century and wireless technology has mostly evolved in other parts of the market. At the same time, the urbanization movement slowed down when many cities became overcrowded and increasingly polluted. Many governments and regional organizations started to work against urbanization by stimulating growth outside the urban areas. Many people, especially young families, moved out of the cities and into smaller communities. Working from home or in local offices became increasingly popular, reducing the amount of traveling as well as traveling time. In the cities, the public transportation systems were modernized and the number of cars decreased significantly, due to strong environmental laws and road taxes when entering city centers. In many cities, large car parks were constructed on the outskirts, which are connected to the city centers by efficient public transportation systems. The result is that fewer people travel long distances to work. Overall this means that daily commuting has increased only slowly since the turn of the century in the industrialized Western world and Japan. The economic recession has also meant that regional and global traveling is at about the same levels as 15 years ago.

One important driver behind this shift in lifestyle is the increasing environmental awareness among the general public. The resulting changes in living, working, and traveling patterns have had an important and negative impact on the use of wireless technology and consequently on the industry. Environmental awareness has, however, affected the wireless industry in other ways too. The increasing need for power to drive computers, routers, server parks (which need heavy cooling), charging battery-powered devices, etc., became an environmental problem during the past decade. When it came to public attention how much of the world's electric energy was used for information and communication purposes, various environmental groups started to campaign for a decrease in the use of communication devices and to put pressure on governments to impose restrictions. For some time, usage was negatively affected but after a few years, industry was able to handle this issue by significantly reducing the power consumption in equipment and devices. The problems with brominated flame retardants (BFRs) also caused a heated debate. Around 2005 it was discovered that large amounts of BFR leaked into groundwater from waste disposal sites where old computer and terminal cases had been dumped. Research showed that these substances could cause children to be born with various defects. In all industrial countries, BFR usage was banned and at some disposal sites these materials had to be dug out and destroyed at a very high cost. These problems also had a significant impact on the use of mobile devices, but soon new and more environmentally friendly materials were developed.

No Service Explosion

Despite the hype at the beginning of the century, the mobile service market never really took off. Instead of exploding, the market experienced slow growth. Most of the services used by consumers are still quite simple, focusing on satisfying basic communication and information needs, such as personal communication through voice, email, and multimedia messaging, searching for information, media broadcasting of news and weather, downloading music, small payments, and simple entertainment services such as games.

Many consumers were simply not prepared to pay for advanced services at the price they were offered. Since a lot of people were accustomed to surfing and using various internet services without high charges, they brought the same mindset into the wireless world. The industry dramatically overestimated consumers' willingness to pay for services other than voice, email, and messaging. It also became evident that the advertising market on the wireless net was nonexistent. Without large user volumes, operators and other service providers were unable to lower the price for communication services quickly enough, leading to a negative spiral. The fear of injuries due to radiation from wireless devices also slowed down the adoption of new services.

Another inhibiting factor was that people were uninterested in walking and surfing. Very few services were compelling enough, considering high tariffs, tiny screens, and primitive keyboards. Most services with high demands on bit rate, quality of service, and security are still used at stationary or portable computers, connected to the internet with fiber or high-capacity wireless access technologies. The wireless killer applications turned out to be voice and messaging—nothing more. Consequently, there was no consumer segment driving the wireless development in terms of new technologies and services. The basic services wanted by the majority of consumers can easily be provided by old GSM networks or updated versions (EDGE, etc.). But this doesn't mean people don't use wireless technologies, it just means their demands are quite like the ones they had around the turn of the century. The two segments least affected by this development have been the Mobile Professionals and Industrial users. Most services are developed for these largely price-insensitive market segments.

For the Mobile Professional, high-quality information services as well as commerce and business services and applications have been developed. Since Mobile Professionals do a lot of globe traveling, it is important for them to be able to connect to home offices to synchronize work and correspondence, but they also use available time to catch up on the latest news from around the world or to plan their coming week. Stock market information is a typical service, together with streaming broadcast news and online ordering of tickets and other goods. Personal communication with family and friends is also important. They use videoconferencing, but almost exclusively as a communication tool for business meetings when fixed conference facilities are not available. Global coverage, seamless communication, high-quality services, personalized information, security, and high data rates are essential for this segment. The terminals they use are powerful and expensive.

The Industrial user segment consists of people with jobs that make them highly mobile and in need of fast and high-capacity data links with their offices and customers, e.g., salespeople, service technicians, couriers, and people providing other logistics services. In this segment, data speed, coverage, positioning, security, tailor-made services, and integration with other systems are of crucial importance. Cost is less important and the function-specific terminals have very high functionality, being a mix of portable and mobile devices.

Wireless Telecommunication is a Mature Industry

The slow growth during the past decade has affected all companies in the telecom industry, especially those focusing on wireless technologies and services. Telecom has become a mature industry that has gone through consolidation and restructuring. The technological development has slowed down considerably and profit margins in all sectors have decreased substantially. Investments in infrastructure have rapidly decreased in the industrialized countries, following the crash about 10 years ago. Almost all surviving companies focus on low-cost systems, terminals, solutions, and services. The importance of brand and customer ownership has increased. Many platforms, solutions, and components are still designed according to closed and incompatible standards protected by patents. Concentration has increased and the number of players in each market is rather few, which means that the surviving players have managed to extend their dominance. Still their profit margins are generally very low or, in many cases, even negative.

The structure of the operator industry has changed dramatically. There are now only a few very large operators, some national but very few small operators, MVNOs, etc. Some operators have survived as publicly subsidized companies, providing coverage in rural areas. Competition is reduced. The European mobile operators that survived the financial problems in the middle of the 2000s managed to keep market dominance by focusing on their customers. The global operators offer services in many countries and have won market share by eliminating high roaming charges when customers stay within their own network. Most operators have split their business into two distinct parts, one focusing on building, running, and maintaining the networks and the other focusing on customer management. By using billing and geographical positioning information from the network, the customer business became a partner for other players, but not very many, building service applications. The network business went wholesale, selling capacity to the customer arm and to other operators and some independent MVNOs.

The system and equipment vendors that survived managed to downsize and focus their business on the large operators. Even though 3G systems have increased in importance for the vendors during the past five years, delivering 2.5G systems and upgrading them is still what provides the bread and butter. Due to the dramatic downturn, no major commercial effort has been made to take the next step after 3G. The big system and equipment vendors have gradually shifted their focus to the NICs, especially China, India, and Russia and to a few developing countries where their business is essentially to provide technology for voice services and SMS. The telco vendors managed to keep most of their market dominance, partly because the large operators are still the most important customers. Their need for integration with the existing systems favors the traditional vendors. Also, most NICs and developing countries with low domestic competence want to rely on the established vendors. The most profitable businesses for the vendors are systems integration and core systems.

The mobile terminal market has been divided into two main segments, one very large and dominating with cheap, simple, and reliable products sold globally at low profit margins. This terminal has become standardized and is a simple device that fulfills basic communication needs. The technology has matured and development and manufacturing is done in low-cost countries. In the other segment, the degree of technological development is higher and these products are tailored for the Mobile Professionals and Industrial users. Prices are high. These very high end products are still developed in Western Europe, the US and Japan, whereas most of the manufacturing is done in low-cost countries. The traditional terminal vendors from the telco industry still dominate the market. Their large share of a market that develops slowly has given them a distinct advantage over attackers from the datacom world. Licensing fees for the software in the wireless device are an important source of revenue, especially for those vendors that have succeeded in establishing their platform as a de facto standard. However, competition is intense from software vendors.

There are very few companies developing applications and services but those that exist are quite large. Their services are standardized and more or less the same wherever in the world they are offered. Almost all the more advanced services are developed and offered to the Mobile Professionals and Industrial users. The price-insensitive professional user, who travels regularly, is perhaps the only segment in which new and advanced generic services are introduced, but they come at a very high price. Among the Industrial users, some companies with a mobile workforce use quite advanced and tailor-made services, but on the whole, the expected increase in efficiency of operations never materialized.

The Big NICs Catching up after a Slow Start

The slow development in the Western world and Japan during the first half of the 2000s was reinforced by problems in the big NICs (China, India, Russia, parts of Southeast Asia, and South America). Even though there were some positive signs in some of these economies around the turn of the century, their difficulties continued. The series of financial crises continued in South America, and Russia was in no better shape around 2005 than it was five years earlier. The integration of China into the world economy slowed down due to political turmoil. After a series of devaluations in Russia, Brazil, and Argentina, civil unrest started, which seriously affected the telecom industry.

However, the upturn came quicker than expected and around 2010 the situation had improved substantially in many of the big NICs. The telecommunication sector has both driven and exploited this development. Investment in infrastructure started to increase, giving the vendors from Europe, Japan, and the US a chance to recover some of their declining sales. China, India, and Russia are now by far the most important markets for the system and terminal vendors. China and Russia have upgraded their old cellular systems (mostly GSM) to 2.5G and are now quickly moving towards 3G. The development in India is somewhat slower, with a current focus on moving from 2G to 2.5G. In all three countries the main focus still is to offer basic communication services (such as voice and simple messaging), in all populated areas, at a price that is affordable to the growing middle class as well as to poorer classes. This has provided an important route to survival for the troubled vendors. In the big cities more advanced services are offered through 2.5G systems complemented with WLAN solutions in hotspots. These services are used by an increasing number of consumers and Mobile Professionals.

Beyond the importance of these countries as markets for the telco industry, they are now bases for important global companies such as operators and vendors as well as service developers and providers. The Indian software industry plays an important role in developing various software products, for use in wireless infrastructure and in devices. The fastest-growing wireless operator with a global scope is based in Shanghai and the most important newcomer in the wireless terminal industry is a Russian company that started to develop and manufacture simple communication devices for the young market segments around the turn of the century. China currently has two infrastructure vendors of global importance. The NICs are gradually catching up with the industrial world.

Spectrum Shortage Not a Big Problem

Due to the radiation problems and the slow development in general, governments in the industrial world have been rather slow in releasing new spectrum. The spectrum released in the infamous 3G auctions in Europe during the first years of the century proved to be more than enough for the demand. Unlicensed spectrum use is limited by very low upper limits on emitted power. Incumbent spectrum holders such as TV broadcasters and the military have been allowed to keep most of their spectrum. Mobile operators still play a dominating role in the industry, since they own most of the spectrum. In the rare cases of new spectrum being released, rules for how to use it were adapted for the traditional operator's business model. Equipment that can be used to bypass spectrum rules, e.g., UWB (ultra wide band), self-configuring base stations, and WLANs, were initially prohibited in the industrialized world in an effort to limit the risks of using it. In the big NICs, especially China, India, and Russia, the governments have released new spectrum as the demand for wireless communication has increased. Spectrum shortage is therefore a minor problem in most countries.

Power Consumption and Complexity Management as Technical Limitations

Even though large amounts of resources have been invested in research into new battery technology, no significant progress has been made. Battery lifetime, for the same size and price, has only doubled since the turn of the century. However, these limitations in the development of batteries did not have a dramatic effect on communication technology in general. The development of processor power, memory size, etc., still largely follows Moore's law. Instead they affected the use of this technology under mobile circumstances, since the performance/weight ratio is still much too unfavorable. Many applications are almost impossible to run when the terminal is on battery power, and even the simple 2.5G handsets and their successors have to be recharged after downloading a few songs or videoclips or after a 1 hour teleconference session.

Industry has tried hard to find ways to minimize the impact of low battery capacity. Some solutions focus mainly on quickly and efficiently replacing or recharging empty batteries. Other solutions involve reducing the power consumption of the device and developing services and applications that need less power. Significant progress has been made in these areas but usage time is still an important limiting factor for most users. Quite a few mobile services have not reached the mass market due to the inconvenience of having to charge or replace the battery too often.

Even though new wireless systems have been deployed at a slow pace, several types of system, based on different technologies, have been introduced during the past decade. In the Western world and Japan, as well as in some of the big NICs, cellular systems of different generations (2G, 2.5G, and 3G) coexist with other types of system (WLANs, PANs, BANs, broadcasting, etc.), some of which are self-configuring whereas others require extensive network planning.

The problems of managing the complexity of this heterogeneous wireless world are still not solved. Many of the systems are isolated islands operated by different entities. The difficulties of achieving integration are massive. Roaming between networks and technologies is expensive and still doesn't offer global coverage. There is a trade-off between the size of the system and the maximum achievable performance in terms of capacity per area and the cost of managing subsystems and databases. The control of the big networks is distributed and the resource management protocols have limited performance; their efficiency decreases dramatically when the size of the network increases. Some ad hoc, self-configurable, and self-healing networks do exist but their size has to be very small. Guaranteeing security and stability in operations is very difficult when the network complexity increases beyond a certain limit.

Wireless Technology in 2015

Still Mostly Second-Generation Wireless Networks

Due to the slow and unsuccessful deployment of 3G, wireless communication is still mainly provided by cellular 2G systems and enhanced versions of them (2.5G). The existing 3G networks were built as islands and are still operated very much in isolation from each other. In many parts of Europe, especially in less populated areas, there is no 3G coverage at all. In most countries there are only one or two different cellular networks and they are operated by traditional telecom operators. Wide area coverage is provided by traditional cellular systems and most of the traffic still goes through these networks. In the US, no spectrum was released for 3G, making 2G/2.5G and EDGE the only cellular networks. In the big NICs, most notably China, India, and Russia, old cellular systems together with enhanced versions also constitute the main wireless infrastructure. Here, however, several new 3G systems are currently being deployed. Consequently, there is still not a single global air interface standard. Just like at the turn of the century, some operators (and nations) have chosen different standards and migration paths, e.g., AMPS, IS-95, GSM/EDGE, WCDMA, and CDMA2000. The old cellular infrastructures offer good QoS in terms of coverage and availability but data rates are low, ranging from a few kbps in rural areas to around 100 kbps in more densely populated areas and cities. Achievable data rates are still limited by restrictions on transmitted power in order to decrease electromagnetic radiation.

In urban areas the cellular systems are complemented by low-power WLANs in hotspots, providing a capacity of 5–20 Mbps in the covered area. But these hotspots are not ubiquitous, and roaming/handover between a hotspot and a cellular system or between hotspots is still expensive and far from seamless. People use hotspots mostly as infostations where data and offline services can be downloaded. Resource management and network control are centralized and the problem of providing seamless integration between various types of air interface is still unresolved. For the cellular systems, traditional network planning techniques dominate. Various broadcast and multicast systems complement the cellular networks by adding more capacity to the downlink. There are also a few satellite systems providing coverage (at low data rates) in very remote areas.

Fear of electromagnetic radiation and public environmental awareness has forced companies to reduce radiation levels (from base stations, etc.) and power consumption (from base stations, switches, routers, and computers) as much as possible. This means that performance has often been sacrificed. In recent years, however, the focus of R&D has gradually shifted. The most immediate problems concerning radiation and power consumption have been solved. Infrastructure vendors have started to focus more on developing low-cost deployment techniques for 3G such as home base stations. In terms of research, industry is mostly focused on solutions for enhancing the 3G systems, whereas universities look further ahead, developing solutions for the next-generation infrastructure. Industry is finally on the verge of introducing 3G systems on a broader scale.

Simple and Low-Radiating Terminals

Most terminals are simple and rather standardized devices with basic functionality, typically voice, messaging, and simple surfing. Many of them are multiband, multistandard terminals, enabling integration of different networks and air interfaces. Together with the large efforts that have been made to cut radiation levels, this has resulted in reduced performance and higher prices for terminals. Mobile Professionals and Industrial users use advanced terminals that are either generic or tailor-made for specific applications. These terminals are expensive and consume large amounts of power, making battery consumption a big problem.

To remove the transmitter as far as possible from the user's head, various BANs are common. Usually the radio transmitter is carried in a bag or sometimes worn on clothing or in the belt. It is connected cordlessly, or sometimes by wire, to an input/output unit worn on the ear or head. For some high-end users, terminal components are built into objects like eyeglasses or belts. Voice recognition and voice synthesis are only working well for the largest global languages. For most small languages, voice control still needs years of development to reach acceptable quality. The same goes for the size of the displays, which most users consider too small to enable advanced services with images and advanced graphics.

Few and Basic Services

Most commonly used are simple communication services such as voice, messaging (email and MMS), location-based information services, broadcasting of news and music, small payments, and simple games. More advanced services come at very high prices and are almost exclusively used by Mobile Professionals and Industrial users. For the common user, the wireless world is anything but transparent and seamless.

4

Rediscovering Harmony

Key phrases: postmaterialistic value shift, balance in life, ad hoc networks, media saturation, environmentalism, fear of radiation, emotional communication, area owners, market refocus

Balance in life became the dominating value in most industrialized nations where material abundance and security could be taken for granted. Media saturation from overexposure to advertising and commercial media promoting a shallow shopping culture triggered a consumer backlash. These are postmaterialistic times where human and environmental needs are in focus, something that is affecting all sectors of society. This development is affecting the wireless industry, which is experiencing a difficult dilemma: refocus or die. There are fewer service and application providers than predicted around 2000, but the market is not completely dry. The mass market is mainly interested in simple peer-to-peer services, messaging services, and various information services. The big hurdle is to refocus and rethink business models, offerings, and brand in relation to a new marketplace with active and demanding consumers categorized by numerous subculture "tribes" with very individual needs and requirements. However, this development is only applicable to industrial countries and markets. In

the NICs, the telco industry is experiencing a more traditional marketplace and business opportunities.

We see a lot of local operators and service providers that have emerged as a result of the trend to move out of the crammed cities and form smaller, local communities where people live and work. At the same time, there are a few global operators providing global communication for the increasing number of people traveling longer and more often for pleasure, and for the smaller but price-insensitive professional and industrial segments. Peer-to-peer services allowing people to trade or exchange digital content make content IPR a hot issue that still needs to be resolved.

Figure 4.1 shows the 14 trends defining the scenario space and how we have set their values for this scenario. The scenario begins with a personal

Figure 4.1 Rediscovering Harmony: defining dimensions. All variables are described in Chapter 6; all user segments are described in Appendix A

Development will be more user driven

Low usage	○ ○ ○ ○ ○ ○ ●	High (moklofs)
Low usage	○ ○ ○ ○ ○ ● ○	High (yupplots)
Low usage	○ ○ ○ ○ ○ ○ ●	High (elders)
Low usage	○ ○ ● ○ ○ ○ ○	High (mobile profes.)
Low usage	○ ○ ● ○ ○ ○ ○	High (industrial users)

User mobility will increase

Low travel growth	○ ● ○ ○ ○ ○ ○	High (cars)
Low travel growth	○ ○ ● ○ ○ ○ ○	High (public transport)

The service and application market will grow

Few services	○ ○ ● ○ ○ ○ ○	Many services

User security, integrity, and privacy will become more important

Security a problem	○ ○ ○ ● ○ ○ ○	Security no problem

Real or perceived health problems will become more important

Radiation a problem	○ ○ ● ○ ○ ○ ○	Radiation no problem

Figure 4.1 (*continued*)

Environmental issues will become more important

Problem   ○ ● ○ ○ ○ ○ ○   No problem

Spectrum will become an increasingly scarce resource

Shortage of spectrum   ○ ○ ○ ⊕ ○ ○ ○   Abundance of spectrum

The wireless industry will grow

Slow growth   ○ ○ ○ ● ○ ○ ○   Fast growth

The big NICs will continue their positive development

Slow growth   ○ ○ ○ ● ○ ○ ○   Fast growth

Market concentration in the wireless industry will change

Few large actors   ○ ○ ○ ● ○ ○ ○   Many actors

The fight for market dominance in the wireless industry will intensify

Operators dominant	○ ○ ○ ○ ● ○ ○	Operators lose
Infra. vendors dominant	○ ○ ○ ● ○ ○ ○	Infra. vendors lose
Term. vendors dominant	○ ○ ○ ● ○ ○ ○	Term. vendors lose

Terminal usage time and complexity management will become problems

Usage time a problem	○ ○ ○ ⊕ ○ ○ ○	Usage time no problem
Complexity a problem	○ ○ ○ ⊕ ○ ○ ○	Complexity no problem

3G will be implemented

Failure   ○ ○ ○ ● ○ ○ ○   Success

Protecting IPR on content will become increasingly difficult

Effective protection   ○ ○ ○ ● ○ ○ ○   Failure of protection

= Not applicable or considered in scenario

story of life in 2015. This story is followed by the main scenario, describing important aspects of the wireless world in 2015. The chapter ends with a brief description of the technical systems and services used.

A Weekday Morning in a Small Scandinavian Village

It's morning and Lisa is woken by her PDA playing a song by her favorite band. It's her best friend, Maria, who has chosen which song Lisa is to wake up to this morning as a surprise, something Maria easily did using an application on her own PDA. The rest of the gang also has this application since it was included with the combined phone/PDAs they all bought together a couple of months ago and today they share calendars, music, and gossip all the time and everywhere using their PDAs.

Lisa quickly checks her PDA for messages in a chat room that she and her friends set up in a public folder in their own wireless hangout last night. Since they are all going backpacking in Nepal later this year, they are now looking for other travelers to meet up with; so far they have found three other groups with similar interests that they will try to get in touch with in different places. There are no new messages, and after checking if her friends are at home or on their way to school, using the Friend Locator positioning service that the group received a week ago when they helped their mobile phone operator to evaluate the software they are currently using, Lisa sends a quick thank you to Maria before hitting the shower.

As usual, Lisa is late for breakfast with her mother and father. After persuading her parents, the family has switched to the more expensive high-quality organic food. Initially, the established agricultural industry ignored this market signal but has now been forced to bow to the consumers. Organic food is now 30% of the total market and growing briskly. At the table they turn on the postcard-sized flat screen and check to see where Lisa's brother Mark is today. He's on a mountain trek in Africa and keeps in touch with the rest of the family using his PDA connected to the family intranet. On the flat screen Lisa and her parents see that Mark has traveled to another village and that he has left a message. "Play message," says Lisa and after a few seconds the screen is filled with Mark's sunburnt face. "Hi there! I'm in Msumba! My phone credits won't allow me to record much, but I just wanted to say that I'm fine and that I hope you guys are okay as well. Lisa, I've uploaded a great song that a South African kid sang last night. Check it out! Take care!" The message is over and the family continue with their break-fast. After a couple of minutes the flat screen flashes and Lisa's grandmother's face shows. "Hello there kids! Nice to see you all up and running, and nice to hear that Mark is having a good time in Africa. I'm going on a picnic with the ladies club today but I'll bring my phone. Can someone check if I've taken my health pills later on and remind me if I've forgotten?"

Grandma was feeling well herself but she wanted to be proactive, just in case. When her doctor, who belonged to the New Total Health Movement, found elevated homocysteine levels, she was warned about the risks of imminent cardiovascular problems and was recommended to start with dietary supplements. Grandma prefers to use cheap natural dietary supplements and now she is taking pills containing high therapeutic levels of vitamins, fatty acids and amino acids. Lisa's father says that he'll remind her and makes a note in his PDA. After talking for a while, grandmother hangs up. The rest of the family finish their breakfast and head off to work and school.

Both Lisa's parents have stopped reading the printed morning papers that they think are too environmentally unfriendly and filled with too much gossip and unnecessary information. They receive their news by belonging to two tailor-made news rings. Lisa's father shares his news and information with his closest friends that more or less have the same interests as him. One of his friends subscribes to a news service that the group receives for free after agreeing to take part in future consumer evaluations, and the information he receives is distributed to the rest of the ring. The news is read using shareware software, which also contains chat software that the friends use to comment on the news and other issues. Lisa's mother gets her news as a service from the company she works for. The company views this as a part of their environmental agenda but also as a way to provide their employees with accurate and corporate-related news.

Lisa takes the school bus and during the trip she uses its WLAN to quickly download Mark's recorded song onto her PDA. The device can't store much data at the moment, but when Lisa and her friends participate in another survey, they'll receive extra memory as a bonus from their mobile operator. She listens to the song and finds it so good that she converts a part of it to a ring signal for her PDA. She uploads the song to the wireless hangout and, just before she arrives at school, she sends a quick message to the rest of the gang about the new signal since today it's Lisa who gets to choose the gang signal. She has to hurry since wireless devices were blocked on the radio-free campus following a community referendum, but she makes it just in time.

Later the same day, Lisa's father's PDA alerts him of grandma's medicine. He opens up the family intranet and clicks on grandma's medical watcher that keeps track of her medicines. He sees that she hasn't taken any pills today and decides to call her. There is no answer. From experience he knows that grandma's PDA is probably covered with blankets in the picnic bag, so he checks the medical watcher, which tells him that grandma's three closest friends are with her on the picnic. To reach her in time, he records a video message that he sends to all of them with a verification button attached. On the way home a couple of minutes later, he receives a message on his PDA saying that one of grandma's friends has verified his message and that everything is okay.

Lisa is on the school bus on her way to track and field training when she receives an alert that her brother Mark is online. She knows that her mother would love to talk to him, so she uses the family intranet wireless portal to set up a group call. After a few minutes the three of them are videoconferencing, Mark and Lisa's mother using computers and webcams, Lisa using the camera on her phone (and the bus WLAN). Lisa's mother even records the conversation for grandma to watch when she gets back from the picnic. After a while, Lisa's father surprises them all by joining in and the family talks a bit longer. Lisa has to stop when she arrives at the training ground and leaves the high-speed network in the bus.

The Wireless Scene in 2015

A Sustainable Society in Balance with Itself

In the late 2000s a slower-paced lifestyle started to influence the Western world. Balance in life became the dominating value in a society where material abundance and security could be taken for granted. In such a society it was just a matter of time before consumers started calling the bluff of marketers and commercial media. This was one of the manifestations of the slow-moving but massive value shifts in the population that had been going on for almost a century. For each new decade, people grew up in a less authoritarian and materially more abundant society. This shifted values from a narrow focus on obedience and material survival to postmaterialistic values. This postmaterialistic world values well-being, tolerance for ambiguity, and individual self-actualization.

The industrialized world of today is based on the idea of a sustainable lifestyle where friends, family, and the environment are key elements. The high-paced lifestyle that dominated the Western world in the closing decades of the twentieth century finally went out of control, creating a society that was running like mad but without any direction. This value shift took many forms. One of the most visible was the rejection of intrusive marketing and the accompanying shopping culture promoted by commercial media, a culture built on creating imaginary and unattainable images about how life should be, breeding media narcissism. Connected to this was an intense public debate on media ethics and media independence. Commercial media were accused of having degenerated into shallow distribution channels for advertisers.

The result of all this was the rise of alternative movements that combined a more sustainable and human perspective on society with a strong individual and social focus. To consider the environment and human needs suddenly became valuable in the marketplace, something that made these movements appealing to more people (as well as businesses and political parties). These movements initially started in the younger segments, but as they gradually became more influential, their values and attitudes started to be reflected in other segments such as the modern businessperson and the Yupplot (young urban person/parent with lack of time).

These values and attitudes are reflected in the political landscape too. In 2015 a large part of the voting population is born in the 1970s or later, making them a politically and commercially important group. Realizing this, political parties quickly started to change their agendas to adapt to the new time, but they also changed as a result of younger party members bringing these values with them into their political work.

Soon another group started to make its voice heard, the Elders. Commercially they are also an important segment with a stable financial situation and a desire to live life to its fullest. However, they are also used to being in charge and they do not accept being disregarded in the course of development.

This shift in values and attitudes hasn't meant that the development of new technologies has come to a halt, but it does mean that the speed and direction of the development have altered. What was once a development of technology based on market-driven issues focusing on hard values such as efficiency, saving time, and commerce, has changed in a softer direction where personal relations, social awareness, and other postmaterialistic values are important. A green wave has reversed the migration waves from the countryside to the cities (or made people stop moving away from their local communities) and has brought with it new possibilities for technological development in more local settings and communities. This is not a hippie movement, but a consequence of political efforts to stop the urbanization trend combined with a reaction against the high-paced society of the early 2000s.

Traveling patterns have now changed as people and companies are suddenly closer to each other. Working from home is a long-awaited success, something that turned out to be a very suitable solution when the distance between home and office was shortened. On the other hand, global traveling has become an important industry since the Moklofs and the Elders are traveling all over the world as often as possible.

This is the truly global society and we are not only global in the way we communicate and travel, but also in the way we think and experience life. Global awareness is natural and issues brought up from the vibrant grass-roots movements and/or from other continents can pop up at any time in the face of companies or in the public debate. However, we are not only part of the global tribe, we are also part of many smaller and local communities and cultures too. The mix between the local and the global ways of living is part of our new lifestyle. The same forces rule the markets, where consumers have started to act globally and unite for or against businesses in various issues, making credibility an extremely important factor. Large companies are very careful not to make exaggerated claims or act unethically. Remember that people today are not against capitalism at heart—they value their own independence too much—but they have started to demand changes using the market's own language. We do not want access to everything all the time, unless it can be achieved through a sustainable and ethical development.

The trend towards a lifestyle focusing on quality of life that we are seeing in the developed countries is not reflected in most NICs. There the urbanization trend continues and the fight for economic gain doesn't seem to stop. However, as these countries gradually increase their quality of living, they are likely to start developing more postmaterialistic values and attitudes.

The Backlash for Marketing and Commercial Media

In the first years of the new century, competition was quickly driving down excess profits in all areas, and in response, companies tried to infuse their products with image, brand loyalty, and celebrity hype. This was one of the few ways left to create higher profits and considerable resources and management attention was devoted to marketing, PR, and media relations. Actual product quality was considered less important than the media image of the same product. When all companies tried to win in the same game, marketing and brand building turned into an arms race. In a world where everybody was screaming, the only way of grabbing attention was to scream louder, be more extreme, or more provocative. The number of commercial messages grew exponentially and in the mid 2000s a lot of the public space was covered with advertising. With growing consumer resistance to advertising, marketers resorted to more indirect schemes such as image building, event marketing, product placement, and the embracing of subcultures. The media, celebrity, and lifestyle industry was the obvious choice for marketers in the attention economy, and the same logic of competition for attention applied in the telecom sector. To draw attention you had to exceed others and be more trendy, sexy, and vulgar.

This was the moment when the general public started to view the prevalent values of the media age—subjective social construction of reality—as production of bluff, chimera, and figment. It was an economy where corporate managements were preoccupied with brand impression while ignoring ethics, environment, human needs, and product quality. When marketing had corrupted language itself into empty fluff, the commercial media were held responsible for enforcing this process and they were accused of having degenerated into shallow distribution channels for marketers. Outrage came when people realized the true extent of how the commercial media agenda was manipulated by PR consultants and spin doctors, hidden cross-promotions and kickback deals.

Market Segments Driving the Development

The move towards a more honest and sustainable lifestyle comes from all segments of society. Two segments can be said to drive this development, the Moklofs and the Elders. The Moklofs are traditionally more interested in family and friends, but today this is more than a passing trend. The Moklofs bring these values with them into their adult lives, even though starting their own families and careers is changing them into Yupplots. The Elders are a segment with quite a lot of power. They are the baby boomers that have become a little older but not quieter. They have high demands on the society they built and they now feel it should take care of them. They want to enjoy the last decades of their lives and they have money to spend. This is also a group that is quite used to modern technology since they have experienced major technological changes during their working years.

The Moklofs are strongly focused on entertainment and messaging services that require less security than many other services. They participate in communities a lot, both local and global, and are very global in their way of thinking. Napster-like services are a perfect example, newsgroups another. Both are also good examples of services that are a bit underground, something that the very postmaterialistic Moklofs have embraced quickly. Content is king for this segment, leading to a vast number of different services and applications. Games and music are popular, but social and environmental services have also turned out to be successful. Always on and always available are two things this segment takes for granted. Though they might value a more human-friendly lifestyle, they require everything at their fingertips, instantly. We are also experiencing a situation where different media types are integrated and interactive. Moklofs are open-minded about new technologies but they don't believe smart marketers trying to claim they will get a new life by buying the latest gizmo.

Living in a world of tribes with many lifestyles, Moklofs want to express their affiliation with clothes, looks, and stuff they use. But after the mad era around 2000 when large cynical companies exploited all subcultures they could find, picking up and destroying their unique attributes by pushing them onto the mass market, the tribes grew extremely wary of being exploited. Most of these tribes have now taken control of their own subcultures and don't tolerate interference of big companies such as MTV. The difference between now and 2000 is that the tribes of 2015 are tribes for themselves and their peers, not a one-way ticket to celebrity and media exposure. Despite a general uneasiness over personal integrity, personal information is often traded to get better services and better deals for this segment. This has led to faster and cheaper development of a variety of simple services; it also means that a credible brand is increasingly important. Users are happy to provide brands they trust with more information about themselves in order to get better services, cheaper fares, more environmentally friendly technology, and so on.

The Elders place a high value on usability. The same goes for the quality of the services they use. They place very high demands on a lot of things, and they are not afraid of letting their voice be heard. Staying in touch with the family while on the move or when living apart has turned out to be very important, creating new contextual and social information technology networks called family intranets. Apart from peer-to-peer communication— keeping in touch with family and friends—healthcare services have become a huge market segment, allowing people to check their health wherever they are. This has highlighted integrity, privacy, and trust, plus the amount of data being transferred. Videoconferencing is an important way of communicating with institutions dealing with personal health or community services. Although technology-friendly, this segment is quite sensitive to the health effects posed by new technologies, real or otherwise.

Although time saving and efficiency are key ideas for the Yupplots, who try their best to combine career with family and friends, the result has turned out to be not only a demand for services that focus on this, but on social services such as being close to friends and family too. Truly mobile services allowing Yupplots to work or maintain social relations while commuting or at home are moderately successful, as well as services that save time, like banking or buying goods on the move. Other, perhaps more surprising, services might be surveillance services that allow working parents to keep track of their kids at daycare or to check up on other household issues. However, the Yupplots are proving to be more stressed by technology than other segments and they are also very concerned about the possible health effects of using mobile communications, especially effects on their children's health.

Less but More Travel

The urbanization movement came to a halt in major parts of the Western world when cities became overcrowded and polluted in the last years of the first decade. At the same time, governments and organizations such as the European Union (EU) decided actively to work against urbanization by providing funding to stimulate growth outside urban areas. People are moving out of crammed cities and into smaller and cleaner local communities in the suburbs or the countryside. The lifestyle trend is that of working and living in small, local, and very social communities.

In the cities, public transportation systems were upgraded and the amount of cars decreased. This was due to harder environmental laws and political decisions to turn more of the city areas into car-free zones or the introduction of high road tariffs, making biking a favorite mode of transportation for many people living a more local lifestyle. Larger car parks were constructed on the outskirts of the cities and connected with the city centers by public transport systems like the subway systems of 2002, but faster. This development is true for the Western world, but not for the NICs. In the NICs, urbanization continues as people try to establish a better quality of life; the result is almost uninhabitable mega-cities marked by overcrowding, pollution, and crime.

Leisure travel is the only form of travel that is increasing all over the world. Two large segments in society, the Moklofs and the Elders, are traveling more than before and further away. The main effect of this is the demand for more environmentally friendly ways to travel, but also the need for global communication possibilities. People want to stay in touch with family and friends, even though they might be Moklofs backpacking in Nepal or Elders golfing in Spain. This has opened up the market for global information and communication providers that coexist with local communication providers. The increased leisure travel has also opened up the market for location-based services on a local scale as well as on a global scale.

A Few Clouds in the Sky

As a result of the development of a more down-to-earth society, a number of issues concerning wireless technologies have emerged. Health risks and integrity problems are widely debated, but it is the telco industry's impact on the environment that people are most concerned with, especially brominated flame retardants (BFRs), used in terminal shells and elsewhere. BFRs have turned out to be damaging to the environment and to humans. Lower power consumption for terminals and infrastructure is something else that consumers want to see.

These issues have affected the telco market in a fundamental way. Large sums of money and resources are being spent on developing less power-consuming technologies and alternative ways of creating power for the networks and the devices connected to them. The same applies to the development of new and less harmful materials used in wireless devices. Discovery of the environmental impacts of wireless technologies has brought with it skepticism from the public. In addressing this skepticism, the telco industry must abide by new rules of honest commitments and it spends huge sums of money on scientific research.

The perceived threats to people's health are harder to counter, forcing the telco industry to find new ways of restoring public trust in its technology. One initiative is an open EU-funded public research project where the telco industry, the World Health Organization (WHO), and local governments are represented. Another initiative is research aimed at developing technologies that produce less radiation; this has been relatively successful and has had a positive effect on the trend of setting up smaller local wireless networks.

The Industry Dilemma: Refocus or Die!

After the financial downturn during the early years of the century, the telecom industry restructured in response to the slow pace of investment in 3G infrastructures. The operators were able to continue their 3G investments when government demands on the pace of the 3G development were reduced and their debts restructured. However, this development depended on the operators generating enough cash flow, and even then the speed of investment in 3G infrastructure was rather slow. Multiband handsets are almost as cheap as simple 2G handsets as a result of rapid technological development. The users were offered a branded 2G/3G service by the operators and found the higher bandwidth and better output from 3G most attractive. This helped to make 3G an established technology, something that gradually made other parts of the world view 3G as the natural standard for wireless infrastructure. The problem of heterogeneous infrastructures was solved by using attractive global roaming tariffs and multiband handsets. This has led to the notion of a single communication network for global travelers, even though they might be in a country that doesn't have a 3G network. Another success factor for 3G compared to WLAN technologies is that the networks offer wide area coverage, something that works hand in hand with the trend to live closer to nature, hence moving further away from other types of communication infrastructure.

After the initial wave of excitement over the new communication possibilities, the pace of development dropped when people and companies found out what kind of technology and services they wanted. This left the telco industry confused. The main reason for this was the industry's inability to adjust to the mass market's new attitudes and values. Many analysts predicted a slowdown in the telco industry around 2010. This turned out to be true in the sense that the pace of change slowed down, but false in the sense that it would lead to a massive crisis. Today the industry is regrouping and adjusting to the new world order where society and the mass market have to be reanalyzed in order to be understood.

The result of this development is that the industry is faced with a serious dilemma: refocus or die. Some players realized this; they adjusted their business models and offerings to the new fragmented marketplace and they are now highly successful. Other companies failed to understand the new environment and are finding it hard to survive. It is clear that there are great opportunities to survive and prosper even in a society dominated by post-materialistic values and attitudes. The big hurdle is to refocus and rethink business models, offerings, and brand. However, this development is only applicable in the industrial countries and markets. In the NICs, the telco industry experiences a more traditional marketplace and well-known business opportunities.

The idea of a brand has been radically redefined over the past decade, yet it remains one of the most important issues for companies. In the new marketplace, brand must be understood as the reward for a total, honest, and credible commitment, not something that you build by media manipulation. Today's market puts enormous pressure on companies to be environmentally aware, to provide tailor-made services and products, as well as to have an ethical vision of their business. Add to this a market that is heavily fragmented with countless subcultures within each existing market segment, each subculture with its own style, values, and attitudes. These subcultures, or tribes, are very sensitive to the behavior of commercial companies, in particular when corporate marketing departments are making noncredible or hypocritical claims.

The more local lifestyle, where people live and work closely, led to the formation of small local clusters of companies and organizations all over suburban areas and the countryside. This turned out to be an excellent space for local service providers and operators that offered companies and people local wireless access and services, often tailor-made for the specific location. On the other hand, the global travel industry has taken off dramatically, making us true citizens of the world. This meant that a number of communication moguls established themselves on the market, providing a global communication infrastructure. We are thus seeing a split market where small and local companies coexist with large global corporations in providing us with a communication infrastructure.

As a result of the relatively slow development of infrastructure after 3G, the equipment vendors are focusing on delivering upgrades to existing systems and on producing cheaper WLAN technology developed in line with the evolution of a more local lifestyle. The result is cheaper and lower-radiating technologies that by no means provide the same coverage as 3G, but which offer local ways of accessing the wireless networks and can be put up by almost anyone.

The terminal market is divided into two main segments: the larger segment produces small, simple, and personalizable wireless devices aimed at Moklofs and Elders; the smaller segment focuses on more costly and powerful devices for Industrial users and Mobile Professionals.

Peer-to-Peer Applications and Services a Hit

Today there is a large market for wireless communication though it looks a little bit different than 10 years ago. Moklofs and Elders, and to some extent

Yupplots, are technology-friendly in terms of using personal (peer-to-peer) communication devices and services, but these services and technologies must be friendly to the environment as well as to people, and they must be easy to use. Keeping in touch with family and friends has turned out to be even more important than expected, and the industry is currently looking for new ways of exploiting this.

We are still information consumers, but not as much as the industry hoped for in the early 2000s, when the information flow peaked dramatically. Society finally came to a point where the saturation of media and the general poor quality of what was being produced made people lose interest and tune out. There is still a demand for information, more or less instantly delivered, but the main difference is that the mass market is selective in terms of what kind of information is being received, and when it is delivered. A new mass-media market gradually emerged, where personalized and very specific types of information service proved successful.

Picture messaging is a massive hit, especially for users in smaller local networks with lower prices for data transfer. These services and applications are simple and easy to use, something that has made them popular in all consumer segments. Another success is personal location-based services that support different types of social activities, such as chatting and keeping track of your friends using simple services and applications, and several types of media such as the internet, digital television, and mobile phones. A third hit is the collection of simple devices that allow parents to check up and communicate with their children at daycare or for families living apart to form simple but emotionally strong family intranets. These family intranets allow families to stay in touch more or less directly, mainly by using simple text and picture messaging.

Another category is directed mainly at Elders who want to maintain personal relations with institutions and organizations in society. This is opening up the market for e-care (medical services). Above all, they need to be trustworthy and easy to use. Videoconferencing using your mobile phone is popular with Elders as it is regarded as more trustworthy to speak to a real person directly. These services require quite a lot of bandwidth, better ways of compressing data, and new ways of making the communication more secure.

The development towards a more local and less hectic lifestyle has made more and more people start telecommuting rather than spending time traveling to and from work. This has created a rather large market for supporting services and technologies. Since these services are mainly used professionally, the main issues are security and the ability to transfer large amounts of data at high speed. Although these services are often accessed through the fixed networks, there is a need for wireless versions too.

Content IPR Still Unresolved

The industry and grass-roots movements drive the development of a new mass-media market. People share information between themselves in small and large clusters of people using Napster-like applications and services, thereby distributing digital content such as news, movie clips, and music on a peer-to-peer basis. Despite generating quite a lot of traffic in the networks, this has had serious implications for the industry, primarily for the content providers that realized they didn't get paid for the information they provided. This has led to a fierce debate on how to solve problems with IPR for digital wireless content. The content providers feel that the operators don't take responsibility for how their networks are being used, whereas the operators argue that they are simply providing the infrastructure for communication. At the same time, consumer groups and governments strongly oppose the idea of surveying the content being distributed in the wireless networks as they see this as a major threat to personal integrity. In attempts to protect their content, the content providers repeatedly come up with new ways of encrypting their information, but every new solution is eventually hacked by underground groups that distribute their work rapidly in the networks. The IPR issue is a problem waiting to be solved.

Wireless Technology in 2015

Many Local and Few Global Wireless Systems

Numerous local as well as a few global companies provide wireless communications and coverage to large parts of the Western world. Parts of the networks are often deployed ad hoc, such as WLANs with data rates up to 100 Mbps or more in the covered hotspot. The lower costs of base stations and access points mean there is also a grass-roots movement of individuals and organizations that put up their own base stations for public access, though with moderate bandwidth and capacity. The local operators are usually affiliated with clearinghouses for automatic handling of access and billing for people entering the area. When they can, local operators act as area owners, trying to restrict competitor access. On a global scale there are relatively few companies that offer multinational communication services based on 2.5G and 3G technologies. High-altitude platforms (HAPs) cover some cities.

All of these systems coexist in a relatively heterogeneous way, mainly because almost all terminals are adaptive and global tariffs are reasonable. This makes the user experience of global communication smooth and ubiquitous, almost as if it is one single system. However, there are still some problems with combining different kinds of air interface, especially between commercial global networks and smaller very local and privately operated networks. Local and global networks require high data speeds for some peer-to-peer services. Another common factor is that they are environmentally friendly (recyclable materials, no harmful radiation) and that they consume little power. Resource management in wireless systems is done by trading available resources. There are several different service levels and users can get better quality for more money. Spectrum is still scarce and is traded between different systems according to their needs; for example, cellular communications borrow bandwidth from TV broadcasters during daytime, when demand is high.

There are two types of terminal: a simple, easy-to-use, and personalized terminal for the mass market and a more powerful and complex terminal for less price-sensitive consumers such as Mobile Professionals and industrial users. A lot of effort is spent on input/output devices such as speech recognition and better screens, and the results are visible. In general, almost all terminals are multimode, adaptive, and easy to interact with.

Simple Services

Most successful services today are peer-to-peer. Picture messaging, personal location-based services, and information sharing (music, video, etc.) are a few examples. These are simple services that are easy to use, allowing people to stay in touch with family and friends using little effort. Telemedicine and e-care is also a big sector of the service market. These services are technically more demanding as they are often based on videoconferencing in real time, putting high demand on real-time compression, stability of transmission, and security.

Standards

There are just a few industry standards and they focus on protecting the big players. Data communication, in wired and wireless networks, is totally

IP-based. Software platforms for the terminals are of two different types: open source and proprietary. Proprietary platforms are preferred for more advanced and very secure devices. Security and IPR are problems that are not yet solved, but several hardware and software solutions have been developed with some success.

5

Big Moguls and Snoopy Governments

Key phrases: market consolidation, few big players, integration, centralized information control, secure services, privacy, priority, reduced competition, winner take all, complexity management

Through consolidation and mergers, large companies, or moguls, have come to dominate the market. A mogul is a descendant of the early big information technology or media companies that managed to survive the crises of the first decade of the twenty-first century. These moguls grew and expanded outside their original business segments; for instance, from being only a systems software manufacturer one company became a big content provider and also started manufacturing devices aimed specifically at its own services. Smaller players were often bought or put out of business due to the dominant position of the big companies. The moguls, together with the world's governments, exert substantial and active control over the information flow and the communication industries. The companies and government are working against the chaotic freedom that used to characterize the early internet, and the purpose is to protect society and individuals from various unwanted actors and behavior. Examples are cybercrime, international terrorism, protecting content owners and others from illegal copying of software, music, movies, etc., and battling other forms of information abuse. The very large and influential moguls are accepted and supported by government since they are seen as more easily monitored than smaller players. Anonymity on the net (fixed and wireless) is no longer possible. All users are automatically identified and registered when acting on the net.

However, the world is not an antidemocratic society where moguls and governments use the net and its information to gain power and ultimately dictatorship, even though many people fear this might be the case. Counter and freedom movements do exist, despite measures against them by governments and large corporations. Development in the big NICs (e.g., China, India, and Russia) has been rather slow during the past decade.

Figure 5.1 shows the 14 trends defining the scenario space and how we have set their values for this scenario. The scenario begins with a personal

Figure 5.1 Big Moguls: defining dimensions. All variables are described in Chapter 6; all user segments are described in Appendix A

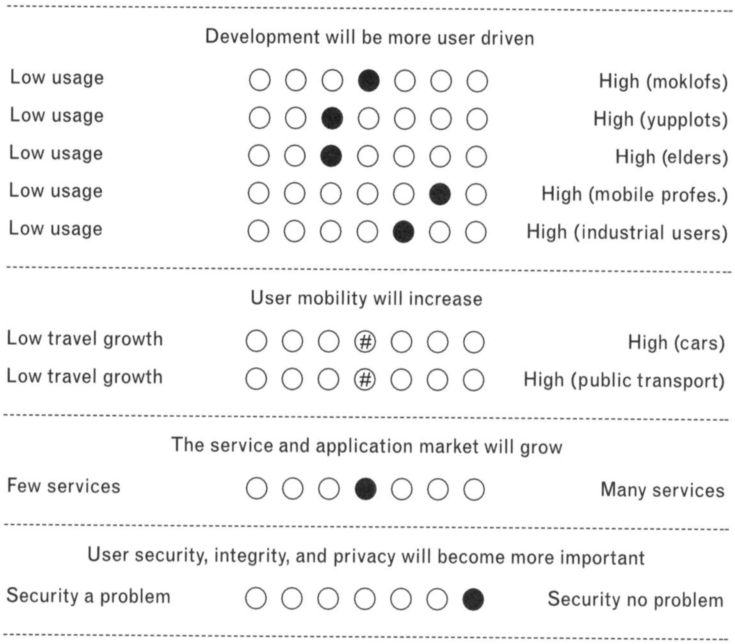

Development will be more user driven

Low usage	○ ○ ○ ● ○ ○ ○	High (moklofs)
Low usage	○ ○ ● ○ ○ ○ ○	High (yupplots)
Low usage	○ ○ ● ○ ○ ○ ○	High (elders)
Low usage	○ ○ ○ ○ ○ ● ○	High (mobile profes.)
Low usage	○ ○ ○ ○ ● ○ ○	High (industrial users)

User mobility will increase

Low travel growth	○ ○ ○ ⊕ ○ ○ ○	High (cars)
Low travel growth	○ ○ ○ ⊕ ○ ○ ○	High (public transport)

The service and application market will grow

Few services	○ ○ ○ ● ○ ○ ○	Many services

User security, integrity, and privacy will become more important

Security a problem	○ ○ ○ ○ ○ ○ ●	Security no problem

Figure 5.1 (*continued*)

Real or perceived health problems will become more important

Radiation a problem   ○ ○ ○ ⊕ ○ ○ ○   Radiation no problem

Environmental issues will become more important

Problem   ○ ○ ○ ⊕ ○ ○ ○   No problem

Spectrum will become an increasingly scarce resource

Shortage of spectrum   ○ ○ ● ○ ○ ○ ○   Abundance of spectrum

The wireless industry will grow

Slow growth   ○ ○ ○ ● ○ ○ ○   Fast growth

The big NICs will continue their positive development

Slow growth   ○ ● ○ ○ ○ ○ ○   Fast growth

Market concentration in the wireless industry will change

Few large actors   ● ○ ○ ○ ○ ○ ○   Many actors

The fight for market dominance in the wireless industry will intensify

Operators dominant   ○ ● ○ ○ ○ ○ ○   Operators lose

Infra. vendors dominant   ○ ● ○ ○ ○ ○ ○   Infra. vendors lose

Term. vendors dominant   ○ ● ○ ○ ○ ○ ○   Term. vendors lose

Terminal usage time and complexity management will become problems

Usage time a problem   ○ ○ ○ ⊕ ○ ○ ○   Usage time no problem

Complexity a problem   ○ ○ ● ○ ○ ○ ○   Complexity no problem

3G will be implemented

Failure   ○ ○ ○ ○ ○ ● ○   Success

Protecting IPR on content will become increasingly difficult

Effective protection   ○ ● ○ ○ ○ ○ ○   Failure of protection

= Not applicable or considered in scenario

story of life in 2015. This story is followed by the main scenario, describing important aspects of the wireless world in 2015. The chapter ends with a brief description of the technical systems and services used.

Early April Morning, Green Haven Gated Community, New York, US

When Ms. Norton sits at her breakfast table this sunny spring morning, she is suddenly struck by the fact that she has now been a very satisfied user of NTN since 2013. "Just like they promised in the ads," she thinks. "Two full years with all our services from the same provider, New Tech Networks, and for a reasonable price." This morning she is reading her usual NTN *New York Times* (paper version, with a nostalgic luxury feel to it), while her five-year-old son and eight-year-old daughter are watching NTNKids on their PlayPads. The boy, who likes "Mr. Jingle's HappyHappyLand" is singing along to the show's theme and the girl is scowling at him because she can't hear her preteen soap, "Barb in School." "If you want to hear use your HappyEar," the boy sings to the tune of the children's show theme, pouting at his sister. "Yes," Ms. Norton says, "both of you use your headsets—all this noise is driving me crazy at breakfast!" They put on their small wireless HappyEar headsets and continue crunching their NTN Brand Cereal, silently focusing on their morning shows.

While the children clean the table, quarreling again, Ms. Norton loads up her day's schedule on the kitchen table display. She works as a government service controller, performing day-to-day management of the local network access points for wired and wireless communication. Holding a degree in wireless systems, she is now senior manager at a nearby facility. Browsing through today's appointments she sees there are some important meetings during the afternoon, but in the morning, just as traffic in the networks is rising, there is little to do except routine tasks. "No one will mind if I do the routine stuff from home and get into the office after lunch," she says to herself. "Then I can run a few machines of laundry in the meantime."

"Let's get the show on the road!" she happily shouts at the kids as they are running late for childcare and school. As they giggle their way to the sport utility vehicle, powered by fuel cells, Ms. Norton clicks on her 3G miniphone and locks the house down through typing a personal identification code and pushing her thumb against the reader on the display. There is a faint hum and hiss as the doors and windows of the house shut themselves tight, and a green light on Ms. Norton's miniphone tells her that the house is now in lockdown mode. It will be safe from any tampering until she or another trusted person tells it otherwise.

"Can we watch NTN on the way, Mum?" the girl asks as they ride out of the suburban community gates. "Sorry, honey," Ms. Norton answers, "I forgot to download during the night, and you know it can't be done while I'm driving." "Why?" asks the boy. "That is so stupid!" "Because of the law," his sister says smugly. "You know it is not allowed to ride in a car and download at the same time. Mum, isn't that so?" "Yes dear," answers Ms. Norton as she waves at Jimmy, the gate guard. "And it is not only because it is dangerous to disturb the driver, it is also very hard to know who is downloading to a car, that's when the car is rolling. When it is connected to a port, like at home, it is easy to see who is doing the download, and where it is being done and that is much, much safer." "I still think it's stupid," the boy mutters sullenly.

When the family arrives at the kindergarten, the boy jumps out, waving his small pad at his sister and mother as he strides up to the high fence. Continuing inside the fence of the kindergarten compound the surveillance camera registers his presence and notifies the preschool teachers that he has arrived. The compound's artificial intelligence (AI) system opens the inner gate and lets him through. At school the guards frisk the girl before letting her in, any electronic devices are to be left at the entrance and the students go to the safe-deposit boxes to get their special school pads, freshly scanned for viruses and forbidden software or media.

Ms. Norton returns home and restarts the house. When she enters, it tells her it has not been tampered with during her absence; it says there are five messages and asks whether she wants to view them now? "Yes," she answers. "Put them on the wallscreen of my study, please." The house does as requested, and after filling the laundry machine and getting a decaf espresso, Ms. Norton strides into her study and begins work.

The messages are all work-related; two are in text-only format and three are multimedia messages. The multimedia messages require her to identify herself to the system using a standard procedure; her 3G miniphone uses the same one. She must enter her 12-digit PIN and a scan of her thumbprint. "It would be nice if it was like this at work as well," she thinks while punching in her PIN. "All that DNA-scanning stuff for logging in isn't really working well yet." While absently pondering security issues, she has a brief look at the multimedia mail and decides it is all red tape that doesn't interest her. "Time to get some work done," she says and starts the commlink to her work facility.

Ms. Norton grew up with normal broadcast television and she is still a bit astounded by the crispness of sound and picture from videoconferencing. The picture she sees is of her office, which she shares with two colleagues. A voice to the left says, "Hi Norton!" It is Ed, her closest colleague, who now shows up on the screen, looking directly at her. "Hi Ed," she says and continues, "How are things over there?" "Well, nothing much happening here," Ed answers. "Rakasha called in sick, but will be available in voice-only." Sharing a smile at Rakasha's vanity, they start going through the night's reports on traffic and picking up a shared window showing the current. "Nothing much, just as I imagined," says Ed. "True," replies Ms. Norton. "How about linking in Rakasha so we can ask her what she knows?" "Okay, here she is."

An icon blinks discreetly at the edge of Ms. Norton's wallscreen, telling her the first laundry is ready. She asks the system to transfer the ongoing conference to her wireless pad, picks up the pad in her hand and walks downstairs to the washing machine. The picture quality deteriorates a little, but the sound quality remains high. Ed blinks owlishly while she puts in another load of dirty laundry. "Hey, nice bathroom!" he says. "Shouldn't you get a wallscreen up here too?"

After lunch Ms. Norton goes out to the sport utility vehicle and drives off to the work, not far from her home. The DNA scanner at the office gates has had some of its bugs ironed out and today's scan is actually quite smooth. She makes a brief stop by her real office, picking up Ed and instantly and wirelessly downloading the upcoming meetings agenda to her secure workpad, which she checked out as soon as the DNA scan was done. Ed picks up his pad as well and tells the stationary system to move his work from the room's wallscreens to the pad. "Oh, I'll have to tell the kids that I might be late for the pickup," Ms. Norton says when she's seen what's going to be covered during the meeting. "At the end we have a discussion of international traffic surveillance policy with France, over link on NTN translation software, and you know how bad that works!" Ed grunts. Then he grins, "Remember when it translated 'chairperson' as 'footstool'?"

Still discussing the pros and cons of automatic translation, they enter the meeting room. The room automatically acknowledges their presence and adjusts their agendas to the latest version. Just as Ms. Norton expected, the vidconf with France is a hassle and takes a long time. Excusing herself for a few minutes, she steps out of the meeting room and picks up her miniphone. "Aw heck, have they radio-shielded the corridor as well now," she mutters when she can't get a signal. Quickly she walks over to her room, where she knows she can make a call. Her calls get routed directly to her kids' pads and on the device screen she can see the boy is deeply involved in some kind of role-play and doesn't seem to be very much bothered by her arriving late. "Sure mum," he says to her. "That's okay. Bye now!" He cuts the link. "He is growing up so fast," she thinks to herself while connecting to her daughter. She can't answer right now as she's attending a sports class, according to her school's AI system, but if Ms. Norton would like to leave a message then that is quite alright. So she does.

Finally the meeting ends and Ms. Norton says a quick goodbye to Ed, reminding him to send over some files tonight. Then she is out in her sport utility vehicle and off to school for the first pickup. When she arrives, her daughter is still inside the building, so Ms. Norton calls the supermarket and orders some burritos and orange for collection in half an hour. She adds the bill to her miniphone account with NTN.

After dinner the kids are off with their PlayPads, watching some shows or interacting with their friends in some of the NTN online worlds. "House, is there anything I'd like to see tonight?" Ms. Norton asks while picking the dirty dishes off the kitchen table. "There is one movie fitting your NTN profile tonight," the house answers. "Shall I put up an excerpt plus comments from your trusted reviewers? Where would you like them displayed?" "Yes please, on the kitchen table." The movie is one she hasn't seen before and it stars some of her favorite actors. The reviewers say it's okay, so she decides to watch it, even though it's late. "Okay, the ban on saving media for later use might be a good way to deter piracy, but sheesh, it can be exhausting trying to stay up just to catch an old film that fifteen years ago I could have taped and watched when I wanted. Then again, the world wasn't so safe in those days." Hearing her boy singing along to NTN's "Learn the Clock with Rock," she decides it would be nice to have some company for the movie, and perhaps the kids would enjoy seeing their father. Smiling to herself, she picks up her miniphone and begins the security procedure.

The Wireless Scene in 2015

Moguls and Governments

Since the mergers of corporations within the IT sector (e.g., media conglomerates, big hardware manufacturers, big software developers, large telecom operators) during 2004–2010, the impact of the big consolidated companies has continued to grow in all markets. In each market segment there are now only one or two totally dominant market leaders. Some market leaders have even been able to expand their market power into other areas. Apart from that, there are few competitors in the market, and those who emerge are soon "brought under the protective wing" of the huge companies. Users like these big dominating companies because they feel they can trust them and their products fulfill their needs. There are no longer compatibility problems with software and hardware as there is only one choice.

Government likes the big companies since they think they can easily control them. To some extent, the moguls agree to this control, as long as the governments are doing what the moguls want. The world's governing forces no longer feel that they need to sell, for instance, spectrum allocation to smaller competitors. The smaller players can be in on the bidding, but they tend to lose to the big players, or else they are soon taken over.

Public acceptance for a more interfering government following terrorist attacks during 2001–2010 led to draconian laws targeting hackers and others who risk committing global or federal (US and EU) crimes. In conjunction with company mergers, new technology for protecting digital media, and tighter cooperation between companies and government, this creates a better, easier, and safer world and has put us where we are today.

As for issues of security and privacy, the big security problems of the early 2000s are almost all solved. Governments work closely together with the dominant companies on policies regarding security and privacy for users of any service. The companies heavily guard their user databases, since much of the market value is based on the concept of trust. Governments have access to almost all user information. Without this integrity for the user, hacking and other IT-related crimes, such as illegal copying of digital media, have become scarce.

On the downside we can see there is only slow development of the wireless industry, unlike in the late 1990s and early 2000s. Since the big companies turned really big, and thus really global, they have no incentives to develop. Governmental and other economic funding for independent research and development is very low. Competition is scarce and pricing can be set to levels that suit the moguls. The economy is stable; interest rates are at a low level.

Security Problems of the 2000s Solved

During the first half of the past decade there was a proliferation in computer-related crime throughout the world. The increasing damage caused by viruses and trojans made computer use an insecure undertaking. The media industry was also heavily burdened by the fact that no media in any digital format was secure from piracy. Starting with peer-to-peer exchange software, where small files were traded, the exploding growth of high bandwidth access and more wirelessly connected devices (hence the possibility of creating ad hoc exchange nets) made it possible to trade freely in any media file. Software and media sales were rapidly dropping.

The media industry tried to stifle their loss of revenue through heavy marketing of events, such as music concerts and through sales of products around their artists, e.g., "the artists brand" clothing. This really proved to be futile in the long run, and if these countermeasures could be taken by the media industry, software companies were not able to hedge their losses from piracy in any way. Intellectual property rights were dying in the globally digital era.

From the early increase in use of internet banks and shops, trust dwindled around 2006, when the encryption standards of the era were growing insecure due to increasingly fast and affordable quantum computers becoming available on the mass market. Newer computers could quite easily number-crunch their way through the encryption algorithms of the time. At the same time, the first mobile device viruses started to spread like wildfire, destroying mobile phones and infecting PDAs. Users started to mistrust the new wireless networks, turning back to barter for security. People reasoned that these problems would remain for some time.

Governments and industries took strong measures against this. In 2007 the first secured devices came out on the market from one of the major hardware and software developers. These devices relied on new, "unbreakable" encryption technologies and required a personal certificate plus user biometrics. They contained circuitry for monitoring traffic and sending information on possibly unapproved traffic directly to the appropriate government agencies.

This set the standard for the market. Users were actually happy to buy and use devices with these functionalities since most reasoned that they had nothing to hide and that they felt secure when doing sensitive transactions like banking. There are still rumors of some havens in the new NICs where those who do not want to show their undertakings to snoopy governments can hide information and carry out shadowy transactions. This is mostly practiced by illicit people, though, and there are strong US and EU governmental movements towards banning these offshore sites.

Moguls in Control

Network effects, economies of scale, and successful enforcing of intellectual property rights created a new global economy with large players becoming even larger, resulting in a winner-take-all society. The US government abandoned the antitrust laws of earlier centuries, allowing already big players from America to grow huge in the truly global market.

As in all previous maturing industries, the IT industry went through consolidation and restructuring. Wireless infrastructure, devices, and services showed very high growth, but they were not new industrial paradigms with new players and new business logic. Hence established players managed to extend their dominance into high-growth markets such as wireless. Wireless growth barely compensated for declining growth in the older parts of the industry, such as traditional telecom and IT. However, globalization and consolidation reduced the number of players in each market, leaving a few winners who had divested all noncore business areas. Markets continued to fragment and in each hyperfocused market niche there were only one or two major market players.

The importance of brand and customer ownership increased in the globalized world, in particular for mobile operators, who got significant market power from controlling networks in many countries. In a maturing industry, mass-market consumers preferred convenience. Incumbent mobile operators managed to offer this and kept market dominance by gradually migrating their huge customer base to new technologies, following government guidelines and schedules.

Over time, network effects on the internet proved to be very strong. Typically, the top three players in each segment got most of the consumers and could leverage and deepen their customer relations. In the wireless area, mobile portals became a convenience handler, a concierge service, and an extension of the person. Over a decade, the leading net players had built deep customer databases and learnt how to use them in a way the customers tolerated. This was typically done by focusing on attractive bundles of services and simple price plans.

Governments released new spectrum in the time period but sold it only to traditional operators; no spectrum was opened to unlicensed use. All previously unlicensed spectrums were repossessed and licensed to the traditional operators. In an effort to crack down on continued terrorism and other crime, governments used a number of measures, such as aggressive IPR enforcement. As a side effect, the media giants became more powerful and this gave additional revenues to the leading software companies. Portals and content sites also benefited.

Slow Development in the NICs

Even though there were quite a few positive signs in the big NICs (e.g., China, India, and Russia) in the early years of the century, their difficulties continued. Financial problems haunted Russia, with frequent devaluations of the currency, and the integration of China into the world economy slowed down due to political instability. The growth of the telecom sector in India has also been slower than expected during the past decade. It is still only the wealthy groups in these countries that can afford anything more advanced than a simple phone. From an industrial perspective this means the markets in these countries are still very important, since the number of people who can afford to spend a substantial amount on telecommunications is large. However, investments in infrastructure have slowed down.

As all the moguls emerged in the industrialized world and strengthened their positions globally, the few emerging players from the NICs were brought under the wings of the moguls in the same way as other small companies. Specialized knowledge was bought, for instance, development of low-powered, high-reliability telephony switching stations for dusty and warm environments (India). Low-cost mobile telephony for developing countries was considered an important issue with a potentially huge market in the early 2000s, but it proved to be too expensive to upgrade the general infrastructure in these countries to a level that would support construction of large-scale wireless communication systems. Western governments keep a keen eye on the NICs, but from a security angle. There have been numerous reports of software and media piracy, havens for pirated media exchange, even talk of antidemocratic movements using the global networks for information exchange. Western governments run intense negotiations with NIC governments, trying to gain access to their networks for surveillance and control of traffic.

Incumbent Telecom Players Keep Control of the Market

With traditional mobile operators dominating over new actors on the wireless markets, the strategic success factors proved to be brand owner-ship and customer relations. The leading European operators managed to survive the financial problems of 2004 through debt restructuring together with government rescue packages and a mild regulatory regime; they have emerged as monopoly players. Weaker operators with financial problems were bought at low prices by the market leaders after they were nationalized by national governments.

After consolidation a few global operators could offer services in many countries and won market share by eliminating high roaming charges when customers stayed within their own network, which became easier when there were fewer network operators to choose from. Customers preferred the convenience of having one provider of all services, and operators integrated them into a seamless branded bundle. Competition drove down traffic prices but the market-leading operators managed to keep market share by following. The volume of wireless voice calls kept growing and gradually replaced landline voice. Together with swiftly growing data traffic, this kept ARPU (average revenue per user) growing in spite of rapidly falling prices. Operators split their business into two halves: one half focused on running the networks and the other half focused on branding, bundling, loyalty programs, and customer management.

The operator/customer company successfully integrated offers from other network providers such as WISPs, ISPs, and fixed access. By using the billing system and geographical positioning information from the network, they became an attractive partner for players interested in building applications for consumers segmented by demography, time, and location. Some operators built successful mobile portals whereas others just provided a platform, selling positioning data and billing with AAA (authentication, authorization, and accounting) to other service providers. The operator/network company went into traffic wholesaling, selling capacity to the customer arm of the operator but also to the remaining independent MVNOs. The rapid technological development of GPRS, 3G, WLANs, global roaming, services, etc., together with technical problems of providing a seamless service, gave traditional mobile operators an advantage over MVNOs. Because mobile access had become a lifeline and extension of the self, most consumers did not want to gamble this on MVNOs, who were obviously not in the wireless industry. The MVNOs were focusing on very price-sensitive segments but were unable to offer more than commodity services. The threat from local WISPs, corporate WLANs, and other forms of do-it-yourself wireless access was handled by the operators in a collaborative way. Operators took the role of clearinghouse, providing local WISPs with a revenue-sharing opportunity while keeping customer ownership of visiting users.

The traditional equipment vendors retained most of their grip on the market. The large global operators continued to dominate as major customers and preferred upgrade paths for their existing infrastructure. Their need for integration with the large investments in legacy systems favored the traditional vendors. As these vendors had established relations with the operators, often going back decades, they were absorbed into the emerging moguls. The large and growing global market gave the incumbent vendors a volume advantage over potential attackers. Most NICs with low domestic competence had to rely on established vendors for their mobile infrastructure.

Operators stopped asking for open APIs and stopped wanting to shop around from several vendors as they folded their own vendors into the organization. Vendors could no longer expect to sign contracts, they simply delivered in-house equipment. Attackers from other markets tried to grab the low end of market niches, and they sparked off corporate warfare. A separate market for plug-and-play modules (e.g., base stations) began to develop, but traditional vendors quickly and forcibly lobbied for legislation to prohibit it.

The large volumes made it possible for the dominating traditional terminal vendors from the cellphone industry to rule the market by scale economics. Users adopted a whole range of terminals: very small voice-controlled units (in glasses or wristwatches), cellphone-sized terminals, and PDA-sized handhelds with a large screen. Most electronic equipment included wireless functionality: laptops, cameras, jewelry, portable games machines, radios, DVD players, and terminals such as goggles with built-in screens. The market was slowly maturing with intense price pressure on the basic terminals. Advanced functions, such as good battery time, high-resolution screens, powerful processors, and small size, are still giving good margins in 2015.

The critical parts of the terminal market became the control software, the silicon processors, batteries, some of the radio interface, and the consumer brand. The software included in all wireless devices was controlled by the few traditional terminal vendors, generating good revenues. Processor makers increased their market power as more and more functionality went into the chip. The market for handheld PDAs grew rapidly but remained a fraction of the cellphone market. Traditional telecom vendors had problems competing with attackers from the datacom industry in this segment but still got revenues by licensing the wireless module and software. Over time, licensing became a significant part of revenues for the telecom vendors. Market power is still in the hands of the traditional cellphone vendors.

3G According to Plan

After the financial crisis around 2004, European governments were forced to create a rescue package for the telecom industry and the mobile operators. Relieved of heavy debts and government demands for rapid 3G investments in rural areas, the operators could generate just enough cash flow to continue their 3G investments at a slower speed. Technological innovations made multiband handsets almost as cheap and simple as 2G-only handsets. Operators offered a branded 2G/3G service and users liked the higher bandwidth and better throughput from 3G. As 3G gradually became an established technology, other parts of the world started to view 3G as a natural alternative for new wireless infrastructure. The multiband handsets and attractive global roaming tariffs from the few existing providers reduced the problem of heterogeneous infrastructure. Global travelers barely noticed that some countries didn't have 3G networks. The share of total traffic in 3G networks grew gradually and 3G is in 2015.

Applications and Services Focus on Convenience for the User

Portals have become very important. Users keep all their information stored at their favorite big company portal, easily accessible from anywhere, at any time. There are numerous applications and services available, but most users prefer the convenience of a one-stop solution. Hence there is a heavy concentration on the top 10 services. As IPR enforcement is quite effective, there are not that many ways of getting "free" content and software, so the users have to stick to the official channels. Users like to be comfortable and want convenience. They trust the well-known brands and are suspicious about new technology. The new moguls' services fulfill their lifestyle needs.

For Moklofs, brand is increasingly important; the users will be happy to provide their favorite brands with more information about themselves in order to get better, newer, or cooler services; cheaper fares; and so on. Some of them delve into offshore sites to form alternative communities, but the mainstream stay with the big companies. The Yupplots think that social issues such as being close to friends and family are becoming increasingly important, together with services that save time or make things more efficient. This means that security and integrity issues are important, something that influences the market for content and service providers, and puts a lot of emphasis on issues such as brand. They are content with the big companies' reliable services and are faithful to their main brand. The 1940s generation are retired and to them brand is important but not only in a commercial way. This segment uses a lot of public services and has a sense of brand in this governmental situation as well. The big companies are trusted and seen as very reliable, having units specifically marketing towards these consumers. The older users seldom change their main service provider.

Mobile Professionals think security is a top priority, together with seamless technologies providing the users with relevant information at the right time. This segment puts the highest demand on the service providers. They are also seen as the most important clients for the companies and are treated with great respect and given customized services. The Mobile Professionals drive the current development to a high extent. Some users have jobs that make them Industrial users more than anything else. These are highly mobile and in need of fast, high-capacity data links with their offices or organizations. The moguls provide them with this, for a price. Like the Mobile Professionals, Industrial users are a driving force of 2015's technological development.

No Free Airwaves

Governments have been very slow to release new spectrum during the past decade. Unlicensed spectrum use is heavily limited by extremely low upper limits of emitted power. As the only spectrum owners for wireless, mobile operators remain the dominant gatekeepers in the industry. If and when new spectrum is released by government, usage rules are adapted for the traditional operator business model. Equipment designed to bypass spectrum rules is prohibited, such as ultra wide band, self-configuring do-it-yourself base stations.

Somewhat of a Complex World

Managing the growing complexity of a varied wireless world has been a problem. With many diverse wireless technologies in the Western world and Japan (such as GSM and iMode) there have been problems of seamless integration between standards and technologies. Multiple generations of cellular systems have been coexisting with other types of wireless connectivity infrastructure (broadcasting, satellites, etc.), causing problems of spectrum allocation. Most of these systems are now proprietary and operated by big companies. The difficulties of achieving integration are mostly solved. Roaming between networks and technologies is expensive but global coverage can almost always be guaranteed. Network operations, for example guaranteeing security and stability, are centrally managed.

Wireless Technology in 2015

Few Different Systems

Few systems, centrally controlled, dominate the wireless world of 2015. The density is high in urban areas and lower in rural areas, but there is almost always coverage. Infrastructure that is centrally owned and managed can be monitored more easily. Bit rate in the wireless networks is often rather low whereas the communication backbone is wired (fiber) and can only be accessed from fixed stations, for security reasons. Thus, the problem of the "last mile," connecting homes to a broadband network, has not been solved through wireless access, but through physically connecting houses at a high cost. It is not possible for the end user to share connections through their own WLAN base station, as these have been prohibited for security reasons. Whenever you access the networks, you are positioned and have to identify yourself.

Global Networks

The cellular networks are global. 3G has taken over globally, totally replacing 2G and 2.5G networks by 2012. Managing the global coverage of the 3G networks is simplified as there are no longer any big issues of roaming between networks and/or virtual operators. Almost all access is centrally controlled and monitored, enabling users to be globally connected and able to get their services wherever they are and whenever they want them.

Since moguls own and control not only land-based infrastructure but also the HAPs (high-altitude platforms) and the LEO (low earth orbit) satellites for communication, the coverage when in very remote areas is guaranteed, but at a cost. These infrastructures also provide access for fast-moving vehicles such as planes and high-speed trains. Personal security of car drivers and pedestrians, and the usual issues of information security mean that in-car systems are only allowed to be used when the vehicle is standing still, connected by wire; this is not very popular. Professional transportation workers download and upload data when stopping, resting, and refueling, as there are fiber connections to their vehicles at strategic points on their routes.

Wireless and Wired Terminals

Terminals for wireless access are not very sophisticated. Since most of the access is from wired terminals, the demand for high-end wireless terminals is small. The main users of wireless high-bandwidth devices are professionals with very specific demands. There is no real mass market for devices that can handle higher bandwidth than that of the 3G networks. The moguls manufacture and sell a rather small range of devices for these networks too; most of them are used for voice communication and text messages (multimedia messaging initially took off but was not found very interesting by the average user).

At home, the terminals in the wired networks are very sophisticated and fairly inexpensive. Moguls subsidize the price of the terminal since the gain in cost for access in the wired networks is high, thus generating more traffic. The home terminal has become the station for all media. Pay-per-view movies, pay-per-listen music, etc., are delivered through the user or user group's portal at their preferred mogul. The mogul is usually the same, owning everything from the fiber to the content.

Usually the mogul also owns the online shops that the user accesses. E-shopping has become appealing since security is more trustworthy now, and the moguls usually keep their promises to deliver the correct goods on time. The home terminal usually consists of a hidden station and an array of multiple big screens, where every user can have their own content. The clunky desktop computer of the last century is long since forgotten. Some devices in the home communicate wirelessly through infrared light and short-range radio.

Quality of Service

The moguls can almost always guarantee available access. If access for certain services is not possible through wireless devices, there are an abundance of public terminals in most urban areas where the identity of the user can be more securely ascertained and high-bandwidth services used.

Few Services but Sophisticated and Popular Services

Since security is very high and personality profiles are stored centrally, users can almost always get what they want. The moguls are fond of complexity and high pricing. It is their strength to develop closed systems. This kind of mogul will always try to lock the customers and end users into their proprietary systems.

Wirelesses devices are used for payment, to get profiled advertisements based on geographical place (e.g., notification of cheap hamburgers when passing a restaurant), secure transactions of money between peers, and so on. For wired communication the bandwidth is very high, enabling streaming of high-quality multimedia and video- or other conferencing. Storing of media is prohibited to fight piracy and governments' intelligent agents travel through networks with almost total access to everything, looking for pirated media in any form. Fines for this crime are substantial.

The moguls' portals are central. Here users buy services and create profiles of their preferences. This is very convenient for the users since they have all their services, from banking to games, in the same place.

Part II

Drivers of Development and Technological Implications

6

Trends and Fundamental Drivers

This chapter gives a theoretical background to trends, fundamental drivers, and theories underlying the scenarios. We have used the trends and drivers to formulate a common ground and frame of reference before developing the actual scenarios.

The four scenarios have been defined by varying the values of 14 trends. These 14 trends form the scenario space and their values are shown in the tables at the beginning of each scenario. Underlying the trends are a number of fundamental drivers, valid in all scenarios. The fundamental drivers are described later in this chapter and we believe they will be valid with reasonably high probability during the next decade. The fundamental drivers are a compilation of common wisdom from a number of areas such as technology, socioeconomics, politics, business, the telecom industry, and user value studies. As the future is always uncertain, they have been formulated as rather vague statements, in order to keep them probable. Some of these drivers are better supported empirically and theoretically than others.

This chapter begins with a description of the 14 trends together with a listing of the drivers underlying each trend. The next section contains a table with the fundamental drivers together with a short description of each driver.

The chapter concludes with an overview of theoretical models, underlying and supporting some of the fundamental drivers.

Fourteen Trends Shaping the Scenarios

The 14 trends in the four scenarios are based on a set of fundamental drivers shaping the development of the wireless world (see later). From the fundamental drivers, 14 trends of particular importance have been identified. These are trends whose direction and rate of change are uncertain. The 14 trends have been used as dimensions when developing the different scenarios. In other words, the scenarios differ along the dimensions of these trends. This section describes the trends and what weight they have been assigned in the scenarios. The descriptions end with a list of the most important fundamental drivers of each trend.

Scenario Abbreviations

- WE is Wireless Explosion—Creative Destruction
- SM is Slow Motion
- RH is Rediscovering Harmony
- BM is Big Moguls and Snoopy Governments

Trend 1: Development Will Be More User Driven

Up to now it can be argued that vendors and technology have driven development within the communication industry. Today there are signs this is changing and that future development within wireless telecommunications will be more user driven. It is, however, not clear how strong this trend is and what user segments will be the most important drivers. Five different user segments have been identified and used in the scenarios: (1) mobile kids with lots of friends (Moklofs), (2) young urban people/parents with lack of time (Yupplots), (3) Elders, (4) Mobile Professionals, and (5) Industrial users (Lindgren, Jedbratt, and Svensson 2001).

The scenarios are differentiated according to the extent to which the development is user driven and according to what segments are the most important drivers in each scenario. WE, for example, is highly user driven with Moklofs and Industrial users as the most important segments. In RH, Moklofs and Elders are most important. In SM, none of the segments are strong drivers except, to some extent, Mobile Professionals and Industrial users (Figure 6.1).

Figure 6.1 Weight of trend 1 in scenarios

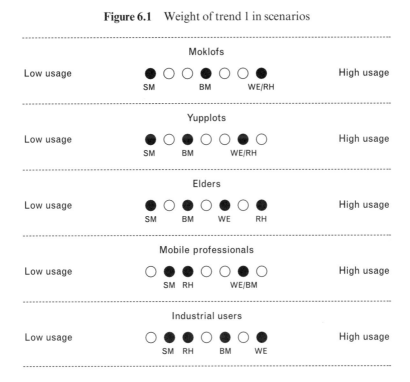

Fundamental drivers

- Need for mobility and communication will increase
- Value of information and knowledge will increase
- Different user groups have different values and needs
- Attractive markets attract new entrants
- Standardization will increase
- Explosion in social group formation in very large networks (Reed's law)

Trend 2: User Mobility Will Increase

For obvious reasons, user mobility is an important aspect for use of wireless services and development of wireless technology. Studies of lifestyle and other related areas indicate that user mobility will increase. We will travel more and longer, both for leisure and professionally. We will also spend

Figure 6.2   Weight of trend 2 in scenarios

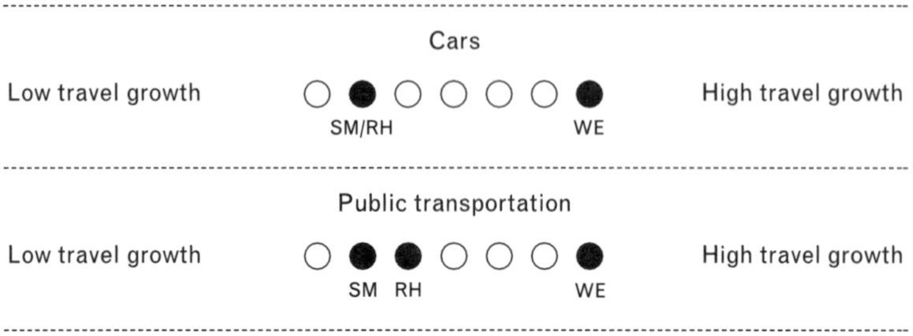

Cars

Low travel growth   ○ ● ○ ○ ○ ●   High travel growth
                       SM/RH   WE

Public transportation

Low travel growth   ○ ● ● ○ ○ ●   High travel growth
                       SM RH   WE

---

more time commuting (IVA 2000). The scenarios are therefore differentiated by how fast traveling will increase and if it will be car travel or travel by public transport that will increase most. Travel is not expected to decrease in any of the scenarios; in WE it increases fast and in SM and RH it increases slowly (Figure 6.2).

Fundamental drivers

• Need for mobility and communication will increase
• Globalization will increase
• Importance of family and friends will increase
• Value of free time and experiences will increase
• Shift towards knowledge industry (OECD countries and NICs)

Trend 3: The Service and Application Market Will Grow

The wireless service and application market will develop from today, essentially only consisting of voice and simple messaging. More advanced types of rudimentary services—surfing, information retrieval, small payments, news, etc.—still only account for a small fraction of the market. The future market will be much larger than today, consisting of more complex and differentiated services as well as the basic ones. The scenarios are differentiated along a service dimension ranging from an abundance of different services and service types in WE to rather few in SM. But even in SM we see a substantial increase of the market compared to today (Figure 6.3).

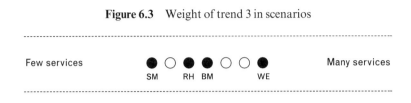

Figure 6.3 Weight of trend 3 in scenarios

Fundamental drivers

- Need for mobility and communication will increase
- Value of information and knowledge will increase
- Different user groups have different values and needs
- Shift towards knowledge industry (OECD countries and NICs)
- Standardization will increase
- Attractive markets attract new entrants
- Globalization will increase

Trend 4: User Security, Integrity, and Privacy Will Become More Important

Guaranteeing security, integrity, and privacy in networks and transmissions is an important problem facing the wireless industry, affecting almost all users as well as companies. The difficulty and complexity of this issue and the unsuccessful efforts to guarantee security on the fixed internet so far, suggest it may not be solved by 2015. The difficulties are even greater in the wireless world, because when you send over radio, in essence you send to everyone. An added complexity is user perception. Even though integrity, privacy, and security cannot be guaranteed, it might be that users don't perceive it as a problem or they just don't care.

This issue is addressed somewhat differently in the different scenarios. In SM consumers are hesitant to use many wireless services for security reasons, since they fear intrusion. It is a problem and most users perceive it as such. In BM this used to be a very big problem, but through interventions from governmental agencies and technological development, security is built into the systems and hence guaranteed. Most users feel secure in using various services, knowing they are protected from intrusions, but at the price of restrictions on allowed behavior on the net, governmental surveillance, etc. In WE the problem itself is not solved at all. The owners of content IPR lose the battle against piracy. Napster-like companies thrive and the users of

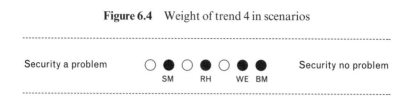

Figure 6.4   Weight of trend 4 in scenarios

wireless services don't care that security cannot be guaranteed. They feel the pros outweigh the cons (Figure 6.4).

Fundamental drivers

• Need for mobility and communication will increase
• Fighting against terrorism and crime, in particular cybercrime, will continue
• Social differences will increase
• Increasing amount of information and choices
• Large and complex systems become increasingly difficult to control centrally
• Increased technology adoption in everyday life
• Explosion in social group formation in very large networks (Reed's law)

Trend 5: Real or Perceived Health Problems Due to Radiation Will Become More Important

One of the biggest potential threats to a positive development of the wireless industry is health problems due to electromagnetic radiation from devices and, to some extent, from base stations and other access points. Today most experts seem to argue that the risks of using cellular devices are very small and all cellular phones on sale radiate at levels well below the allowed SAR value (SNRV 2002). However, further investigations have to be performed on for example long-term effects of radiation. It might very well be that such studies will show an increased risk of developing brain damage and cancer due to radiation, like in SM. Moreover, the fact that many people worry about health effects, whether real or perceived, makes this a very real problem that needs to be handled by the wireless industry. The scenarios are differentiated according to how big a problem the health issues (real or perceived) relating to electromagnetic radiation will be (Figure 6.5).

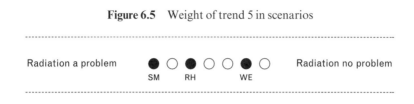

Figure 6.5 Weight of trend 5 in scenarios

Fundamental drivers

- Environmentalism and health concerns will increase
- Need for mobility and communication will increase
- Increased technology adoption in everyday life

Trend 6: Environmental Issues Will Become More Important

Environmental awareness in society has increased during the past decade. It seems clear that this trend will continue, forcing industry to develop increasingly environmentally friendly products. This will affect the wireless industry. There are at least two areas of special importance: energy consumption and potentially detrimental substances used in terminals and other equipment.

The global need for electric energy for the net and the wireless systems is not negligible compared to other sectors of society and industry. Each server park already consumes power in the megawatt range; add to this hundreds of millions of battery chargers, PCs, and base stations and the problem becomes significant. Environmental groups or governments might begin to focus their attention on the telecommunications industry, forcing companies to develop more energy-efficient systems and products, like in SM. Current research indicates that several substances, especially some brominated flame retardants (BFRs) used in terminal cases, might be dangerous to ecosystems (human beings as well as animals and plants) if they leak from waste disposal sites. The scenarios differ in the size of this problem (Figure 6.6).

Figure 6.6 Weight of trend 6 in scenarios

Environmental issues a problem	○ ● ○ ○ ○ ● ○	Environmental issues no problem
	SM/RH WE	

Fundamental drivers

- Environmentalism and health concerns will increase
- Need for mobility and communication will increase
- Increased technology adoption in everyday life
- Processing power will increase exponentially

Trend 7: Spectrum Will Become an Increasingly Scarce Resource

Spectrum is a resource of great importance for wireless industrial development. For historical reasons, most spectrum is locked in by legacy users, the military and television broadcasters. Wireless consumer communications has been given less than one-tenth of all usable spectrum between 500 MHz and 5 GHz. In Europe the military on average controls 30% of all spectrum. In combination with the rapid increase of wireless users during the past decade, this has caused a shortage of spectrum in many areas. This shortage is forcing the operators to build unnecessary and expensive infrastructure with many base stations and is today a real bottleneck in many cities. Growing usage will only aggravate this problem, and spectrum shortage is potentially one of the biggest inhibitors to further growth. If governments allocate abundant spectrum to the wireless industry, growth will be much faster and prices for the consumer will be lower. If governments are slow to release new spectrum, growth will be slowed. Spectrum can be released either to the operators or to unlicensed bands for free use. Unlicensed spectrum will undermine the market power of the operators and create a more dynamic, but anarchistic and less controllable marketplace. Another way to reduce the problem is to develop technology, enabling a more efficient use of the existing spectrum.

In WE spectrum shortage is not a problem since governments release more spectrum when needed, at the same time as technological development improves spectrum efficiency of the systems. Spectrum is released to operators and for unlicensed use. In BM governments are slow to release new spectrum, hampering the development of industry. No spectrum is released to unlicensed bands for security reasons (Figure 6.7).

Fundamental drivers

- Air bandwidth is affected by political decisions
- Capacity in air will increase but slower than in fiber

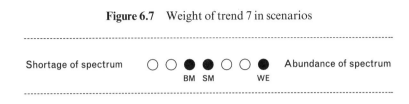

Figure 6.7 Weight of trend 7 in scenarios

- Processing power will increase exponentially
- Need for mobility and communication will increase
- Increasing amount of information and choices
- Increased technology adoption in everyday life

Trend 8: The Wireless Industry Will Grow

All scenarios are based on the assumption that the wireless communications industry, i.e., operators, vendors of infrastructure and terminals, and service developers and providers, will grow during the coming decade, both in size and scope. The question is how fast this growth will be. In WE industry grows very rapidly, whereas it grows very slowly in SM (Figure 6.8).

Figure 6.8 Weight of trend 8 in scenarios

Slow growth ●○○●○○● Fast growth
SM   RH/BM   WE

Fundamental drivers

- Need for mobility and communication will increase
- Globalization will increase
- Shift towards knowledge industry (OECD countries and NICs)
- Market economy prevails
- Attractive markets attract new entrants
- Increased technology adoption in everyday life
- Standardization will increase
- Scale and learning economics (price/performance)
- Industries mature over time (negative impact)
- Complexity diseconomics (negative impact)

Trend 9: The Big NICs Will Continue Their Positive Development

There are many signs of positive development of the telecommunications sector and especially the wireless sector, in the most important NICs, e.g., China and others (Long 2002; Bhattacharya 2002; Schildknecht 2002). The markets in these countries are enormous and the growth rates are high. New companies are established, especially in China, with an ambition of becoming global players for example, as infrastructure vendors. The big NICs might become important not only as big markets for European, American, and Japanese companies, but also as countries where important new competitors emerge (Figure 6.9).

Figure 6.9   Weight of trend 9 in scenarios

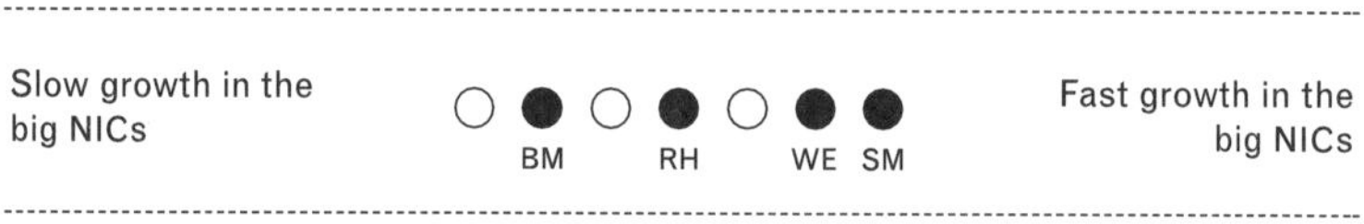

Slow growth in the big NICs   ○ ● ○ ● ○ ● ●   Fast growth in the big NICs
                                                BM   RH   WE SM

Fundamental drivers

• Globalization will increase
• Democratization will increase
• Market economy prevails
• Shift towards knowledge industry (OECD countries and NICs)
• Need for mobility and communication will increase
• Social differences will increase (negative impact)
• Scale and learning economics (price/performance, negative impact)
• Industries mature over time (negative impact)

Trend 10: Market Concentration in the Wireless Industry Will Change

The future structure of the wireless industry is an open issue. Several factors are pointing towards increased concentration with a few market leaders wielding great market power, but other factors are pointing towards a fragmented marketplace where the market leaders have little market power. It is possible that the industry (or certain parts of it) is facing a consolidation phase, with fewer but bigger actors. This is the case in BM and SM. In WE

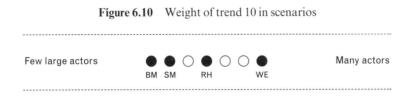

Figure 6.10 Weight of trend 10 in scenarios

the number of actors increases in all sectors of industry. The market leaders are relatively weak and are unable to wield any market power. There are a large number of companies and many different types of companies competing on a market with very narrow business models (Figure 6.10).

Fundamental drivers

- Industries mature over time
- Companies strive towards monopoly
- Scale and learning economics (price/perform ance)
- Attractive markets attract new entrants
- Value chains will increase in complexity (value networks)
- Standardization will increase
- Globalization will increase

Trend 11: The Fight for Market Dominance in the Wireless Industry Will Intensify

The merging of telecommunication, data communication, and media into an at least partly integrated industry will have an important impact on the existing telecommunications companies. Telco infrastructure and terminal vendors are increasingly threatened by companies from the datacom industry, e.g., vendors of base stations and access points, routers, PDAs, laptop computers, and software. It is not at all clear which industry will emerge as the long-term winner. Operators might lose market power to new WISPs or through media companies owning content being integrated with wireless services. In WE the traditional telco vendors lose to attackers from the datacom world. Operators also lose dominance to new players. In SM and BM the traditional telcos sustain their industry dominance, but for different reasons (Figure 6.11).

Figure 6.11   Weight of trend 11 in scenarios

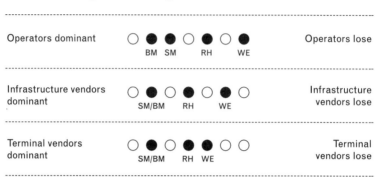

Fundamental drivers

- Industries mature over time
- Companies strive towards monopoly
- Attractive markets attract new entrants
- Scale and learning economics (price/performance)
- Standardization will increase
- Globalization will increase
- Internet development dominating
- Large and complex systems become increasingly difficult to control centrally

Trend 12: Short Terminal Usage Time and Complexity Management Will Become Increasingly Important Problems

There are many potentially important technical problems that might have a negative impact on the development of the wireless industry. Two of these are power consumption in the mobile device and the problem of simultaneously managing many complex and heterogeneous wireless systems. From a user's viewpoint, usage and standby time of the wireless device are crucial. With the rapid increase of processor power and memory size, the much slower development of battery capacity will make usage and standby time a problem. In WE this problem has been solved, but it still remains in SM. The same goes for managing the complexity of a heterogeneous wireless world. The problems of integrating various wireless

Figure 6.12   Weight of trend 12 in scenarios

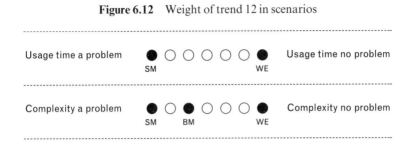

systems, achieving seamless services from a user's viewpoint, are solved in WE but not in SM (Figure 6.12).

Fundamental drivers

- Processing power will increase exponentially (usage time)
- Fiber and memory capacity will increase exponentially (usage time)
- Capacity in air will increase but slower than in fiber (usage time)
- Battery capacity will increase very slowly (usage time)
- Need for mobility and communication will increase (power and complexity)
- Complexity diseconomics (complexity)
- Large and complex systems become increasingly difficult to control centrally (complexity)

Trend 13: 3G Will Be Implemented

Currently one of the most important issues for the wireless industry, for operators as well as vendors, is the deployment of 3G. It seems clear that 3G will be implemented, but the question is at what speed and to what extent, with operators currently burdened by extreme debts. In SM, 3G is a failure. It is deployed later than scheduled and many networks are canceled or merged with other networks, leading to a spotty situation with barely one physical network per country and no 3G coverage in large parts of rural Europe. In the US no new spectrum is allocated to 3G. In BM, on the other hand, 3G is built almost according to plan, becoming quite a success (Figure 6.13).

Figure 6.13   Weight of trend 13 in scenarios

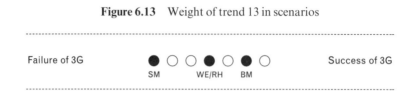

Failure of 3G   ● ○ ○ ● ○ ● ○   Success of 3G
                         SM   WE/RH BM

Fundamental drivers

- Wireless infrastructure cost will fall slower than for electronics
- Environmentalism and health concerns will increase
- Air bandwidth is affected by political decisions
- Need for mobility and communication will increase
- Complexity diseconomics

Trend 14: Protecting IPR on Content Will Become Increasingly Difficult

The problem of protecting intellectual property rights (IPR), especially on content offered to wireless users, is very important for industry. Many new services will depend on delivery of content owned by someone other than the user or the service provider. Will the Napster-like development of today continue or will the content owners, e.g., media companies, be able to protect their content and share profits with the service providers? In the BM scenario, IPR enforcement succeeds but at quite a high price. In WE an anarchistic wireless world develops where piracy prevails and much protected content is spread to users beyond the reach of the content owners (Figure 6.14).

Figure 6.14   Weight of trend 14 in scenarios

Effective protection   ○ ● ○ ● ○ ○ ●   Failure of protection
                                BM   RH   WE

Fundamental drivers

- Digitalization will increase
- Internet development dominating
- Fighting against terrorism and crime, in particular cybercrime, will continue
- Explosion in social group formation in very large networks (Reed's law)
- Increasing amount of information and choices

Fundamental Drivers

Underlying the trends used as dimensions to define the scenario space are a number of fundamental drivers, valid in all scenarios. The fundamental drivers can be found in Table 6.1 and we believe they will be valid with reasonably high probability in the next decade. The drivers are "mega-trends", a compilation of common wisdom from a number of areas, such as technology, socioeconomics, politics, business, the telecom industry, and user values. As the future is always uncertain, they have been formulated as rather vague statements in order to keep them probable. Some of these drivers are better supported empirically and theoretically than others. There are a number of fundamental tenets in technoeconomic development that can almost be formulated as laws when considering their empirical support.

Technology Drivers

Processing power will increase exponentially (Moore's law)

During the period until 2015, processor capacity will continue to increase by 55–59% annually, doubling every 18 months.

Fiber and memory capacity will increase exponentially

Memory capacity is growing exponentially in the same range as processor capacity. Optical fiber capacity is growing faster, about 90% per year (doubling every 13 months).

Capacity in air will increase but slower than in fiber

Capacity in air is the data capacity in a given geographical area. It follows from spectral efficiency, which denotes how much data can be transferred

Table 6.1 Fundamental drivers of development

Technology	Socioeconomic and political	Business and industry	Users, values, attitudes
Processing power will increase exponentially (Moore's law)	Globalization will increase	Industries mature over time	Values change at the pace of generations
Fiber and memory capacity will increase exponentially	Democratization will increase	Companies strive towards monopoly	Individualism will increase
Capacity in air will increase but slower than in fiber	Aging population (Industrial world)	Attractive markets attract new entrants	Importance of family and friends will increase
Battery capacity will increase very slowly	Shift towards knowledge industry (OECD countries and NICs)	Scale and learning economics (improving price/performance)	Value of free time and experiences will increase
Miniaturization of components will increase	Market economy prevails (but counter movements will continue)	Complexity diseconomics	Need for mobility and communication will increase
Digitalization will increase	Fight against terrorism and crime, in particular cybercrime, will continue	Value chains will increase in complexity (value networks)	Increased technology adoption in everyday life
Standardization will increase	Social differences will increase	Wireless infrastructure cost will fall more slowly than for electronics (Zander's law)	Different user groups have different values and needs
Internet development dominating	Increasing amount of information and choices	Value of network increases with number of nodes (Metcalfe's law)	Environmentalism and health concerns will increase
Large and complex systems become increasingly difficult to control centrally	Air bandwidth is affected by political decisions	Value of information and knowledge will increase	Explosion in social group formation in very large networks (Reed's law)

in a given small slot of spectrum. According to Shannon's law, spectral efficiency has an upper limit and we are now approaching this limit. To keep capacity growing, we need to cover smaller areas with access points or use smart directional antennas.

Battery capacity will increase very slowly

The low energy density in batteries is one of the big bottlenecks in the wireless industry. Capacity will grow in the next decade but not at a pace that will give future power-hungry devices lightweight batteries with a long lifetime under heavy use.

Miniaturization of components will increase

Miniaturization is an alternative way of exploiting the increased performance of components. With smaller and more powerful components it will be possible to make smaller devices.

Digitalization will increase

The trend in IT and telecom of moving from analog to digital will continue.

Standardization will increase

Modularization and open APIs is a trend that will continue. Software developers and industry are very skeptical about the risk of being locked into proprietary solutions and are asking for open standards.

Internet development will be dominating

There are no signs that the mega-trend of internet will come to a halt before 2015. Internet Protocol (IP) will totally dominate wireless. Eventually something else will replace the internet but this is still way beyond the horizon.

Large and complex systems become increasingly difficult to control centrally

In a world of ubiquitous computing and device proliferation, the old telco-style world with centrally controlled systems will be unmanageable due to complexity problems. Other distributed alternatives have to be used to solve the problems.

Socioeconomic and Political Drivers

Globalization will increase

Increasing globalization follows from the century-old trend of falling costs of transportation and communication. Global trade and movement of people and ideas are increasing. These are long-term mega-trends and have been in place for centuries. War and political backlash can temporarily reverse this trend but in the long run it is inexorable. Travel and communication are the enablers of increasing globalization.

Democratization will increase

For decades the number of democratic countries has increased; very few democratic countries have fallen back and stayed undemocratic in recent decades. The ideological winner is obviously democracy.

Aging population (industrial world)

Demographic data show falling birth rates and that the older segment of the population will keep growing.

Shift towards knowledge industry (OECD countries and NICs)

The service and knowledge economy has already replaced industrial manufacturing as the major economic driver in the OECD countries. The knowledge economy is now a growing sector in the NICs too.

Market economy prevails but counter movements will continue

The global market economy is stronger than ever and has few serious challengers. However, counter movements will continue and focus on specific issues and problems in the market system.

Fighting against terrorism and crime, in particular cybercrime will continue

In a globalized world, law enforcement still operates in a world of national states. Law enforcement will increase global cooperation to meet the threats from global crime, terrorism and borderless cybercrime.

Social differences will increase

In a rapidly changing society the differences between those who can benefit from the changes and those left behind will be large and very visible.

Increasing amount of information and choices

The net is opening a world of infinite access to information. This open architecture will be dominating in the wireless world too.

Air bandwidth is affected by political decisions

Governments control spectrum usage. Decisions to release new spectrum to the wireless industry could considerably increase available air bandwidth.

Business and Industry Drivers

Industries mature over time

Industries created by technological innovations form an S-shaped life cycle, sometimes spanning up to a century. The IT industry is now over 50 years old and the fixed telecom industry is entering its second century. Eventually the wireless industry will begin to transform into a mature industry. The question is when this will happen.

Companies strive towards monopoly

The best way to achieve high profits is to lock customers in and competitors out, i.e., to create a temporary monopoly. In spite of rhetorical commitment to principles of open platforms and standards, a company with strong technology or market power will try to expand its position and become a new Microsoft-type monopolist.

Attractive markets attract new entrants

High profits and a rapidly growing market are like a magnet, attracting many players to enter. This will increase competition and eventually drive profits down.

Scale and learning economics (improving price/performance)

Larger production volumes give falling costs. The experience curve states that for each doubling of accumulated production volume, cost falls between 25% and 35%.

Complexity diseconomics

In a fast-moving environment, it will be increasingly hard for large integrated companies to keep up. To survive in the market, companies must concentrate on the market segment where they are most competitive. The large incumbent telcos are particularly vulnerable to complexity diseconomics.

Value chains will increase in complexity (value networks)

A larger market can sustain more specialized business models. The linear value chains in the industrial economy are replaced by much more complex web-like relations in the information and knowledge economy. When each company is focusing on its own niche, it needs many more external relations.

Wireless infrastructure costs will fall slower than electronics costs
(Zander's law)

Scale economics and Moore's law have been radically driving down the price of electronic equipment during the past decade. This does not apply to the construction costs for masts and cabling. The cost for a large 3G mast is now 80% construction and only 20% electronics.

Value of network increases with number of nodes (Metcalfe's law)

Total network value rises by the square of the number of terminals.

Value of information and knowledge will increase

In the postindustrial knowledge-based economy, knowledge and information are the critical assets.

Users, Values, and Attitude Drivers

Values change at the pace of generations

The fundamental values a young person acquires before age 25 usually remain unchanged for the rest of his or her life. This considerably slows the rate of value shifts in society. It is only by new generations replacing the earlier generations that fundamental values change can occur.

Individualism will increase

This is one of the most important value shifts that has been observed. People born in any given decade have more individualistic values than people born in the previous decade.

Importance of family and friends will increase

This is another important value shift. People are more individualistic yet very concerned about forming social networks with family, friends, and other groups.

Value of free time and experiences will increase

Increased material abundance brings a value shift from materialistic values to postmaterialistic values. With most basic material needs fulfilled, people will change their priorities away from making more money to quality of life. This is less pronounced in the NICs, where striving for material gains will dominate the value system for the next generation.

Need for mobility and communication will increase

People will live a more mobile lifestyle, which will increase the demand for communication and personal mobility.

Increased technology adoption in everyday life

Technology is invading our daily lives. This long-term trend will continue and speed up with the proliferation of new devices.

Different user groups have different values and needs

In a society with increasing social differences and diversity in lifestyles and values, it will be very obvious that the needs are very different between various user groups.

Environmentalism and health concerns will increase

Value studies show that environmental responsibility is taken for granted by the postmaterialist younger part of the population. With increased individualism comes a sense of being responsible for one's own life and falling respect for authorities. One consequence is that perceived health hazards, such as cellphone radiation, will be a major concern.

Explosion in social group formation in very large networks (Reed's law)

In networks such as email and the web, people can form an almost infinite number of groups among themselves. The number of possible groups grows exponentially with network size. When networks grow very large, this effect is much more important than the value of connectivity (Metcalfe's law), growing as only the square of network size.

Theories Supporting Fundamental Drivers

The scenarios and the fundamental drivers are based on models concerning value shifts, technological innovations, and industrial development. Here are a selection of theories that we believe are the most important.

Exponential Growth

One of the most fundamental features is that technological development and economic development often follow exponential growth curves. This tenet is often overlooked as human attention is usually focused on the immediate present, past, and future. Exponential growth curves look linear in a short-term perspective, accentuated by the common use of linear diagrams, which makes us think in terms of linear change. The consequence is that people often underrate the dramatic development in the long term. That is, exponential growth following Moore's law is not an exception but just a very visible case of rapid growth. In the tech industry, growth can sometimes be even faster than Moore's law; in the industrial economy, it can be much slower but still exponential. These exponential curves show up everywhere: in price/performance, in reduced emissions, in falling prices, in higher capacity, in reduced raw material input for production, etc. When one exponential growth path reaches maturity and slows down, new technologies are ready to take over. Newer technologies usually grow faster than the older technologies they are replacing. In the early twentieth century, quality-adjusted US car prices fell by 5% per year during 1906–1940, when cars were the cutting edge of high technology. Half a century later, quality-adjusted PC prices fell by about 20–30% per year during 1982–1988 (Raff and Trajtenberg 1997).

Even low growth figures give remarkable gains over long periods of time. The price of illumination has been falling by 3.6% per year for 200 years. This exponential path has been valid through the shift from candles to town gas lamps to kerosene lamps to electric light bulbs to fluorescent bulbs. The real price today for one lumen-hour is below 0.1% of what people had to pay 200 years ago (Nordhaus 1997).

These long-term price falls go unnoticed by most people. Preoccupation with new gizmos hides the fact that most necessities are becoming so cheap and of such high quality that they fall below the mental radar. Expensive new products that consumers now desire didn't exist a few years ago. On aggregate, these growth paths drive economic and GDP growth, themselves following exponential paths. If we assume a growth in GDP per capita of just 2.8%, in 50 years we will be four times better off than today. And the GDP measure vastly underestimates how much better off we will be. GDP doesn't capture increased quality, more convenience, and the benefits of totally new products and services solving needs previously impossible to fulfill.

Microprocessor and Other Growth Paths

In the wireless industry we assume that Moore's law will be valid until 2015 and the growth paths in the same range for memory storage capacity and fiber capacity will also hold until 2015. The exponential growth in performance has been valid over a number of technology generations, and from 1940 computing power has grown steadily by 55% (Nordhaus 2001), closely matching the 59% performance growth of silicon processors stated as Moore's law. Battery capacity is also growing but much slower than Moore's law. Capacity in air (defined as spectral efficiency/covered area) is another much slower but still exponential growth path.

Exponentially Falling Prices and the Experience Curve

When new technologies reach the market, economies of scale will cause prices to fall as production volumes increase. This trend of falling prices is called the learning curve or the experience curve and was developed by the founder of the Boston Consulting Group in the 1960s. It has been empirically shown that for each doubling of accumulated production volume, prices will fall 25–35%. The experience curve has been shown to hold for such diverse products as photovoltaic solar cells, B17 bombers during World War II, and AT&T long-distance voice tariffs over a 50-year period. (Grübler 1998; Mishina 1999).

Network Effects I (Metcalfe's Law)

When a successful new technology gets a large user base, network effects will give additional benefits for each extra user. For example, the more people getting a fax machine, the better for everybody else with a fax, as they will be able to reach more people. This case has been stated as Metcalfe's law—total network value rises by the square of the number of terminals (i.e., by N^2). Network effects don't only appear in networks. In software, network effects also arise from a large installed base, which will enforce the market leadership when the dominant software gets a large base of software developers (Shapiro and Varian 1998). These network effects are very important in the early market when several technologies are competing with each other. When one technology gets a lead by a large market share, this advantage is often enough to tip the balance and enforce market leadership. This transition is enforcing one dominant design, and once that is in place, the technology is ready to explode on the mass market. In the software industry, this transition has been called crossing the chasm and the ensuing explosion in sales and market size is called the tornado (Moore 1991, 1995).

Network Effects II (Reed's Law)

Metcalfe's law is a valid model for one-to-one communication, where you call or fax another subscriber on the network. The network effects of Metcalfe's law are impressive but the really large effects will be found when looking at the number of groups that can be formed among the network members. Reed's law (Reed 1999) states that in a given network of size N, the number of nontrivial groups that can be formed grows exponentially, i.e., as

2^N. (The number of nontrivial subsets that can be formed from a group of N members is $2^N - N - 1$.) Networks that support communicating groups, Reed calls them group-forming networks (GFNs). The web and email are GFNs but the traditional telephone network is not a GFN. When N becomes large, the term 2^N will grow much faster than the term N^2 in Metcalfe's law. Even if the value of connectivity (expressed by Metcalfe's law) is much larger than the value of group formation, when the network grows large, the GFN effect will dominate. This is an indication of the potentially very large impact on social life and community building when the global network grows very large.

The S-curve and the Product Life Cycle

Successful new technologies follow an S-curve when penetrating the market. During their life cycle, new technologies have different categories as buyers. New technologies will first find buyers among innovators, followed by early adopters, early majority, late majority, and finally laggards. Adoption of the new technology (the slope of the S-curve) is fastest in the middle, when the S-curve is halfway towards maturity.

The product life cycle is basically a model using the S-curve but with focus on the producer. For the producer, growth and profits are strongest in the phase when the early majority adopts the new product. Profitability is usually falling dramatically when the product matures but the very large total market in the late phase of the S-curve keeps absolute profit figures reasonably high. Eventually the product reaches maturity, saturation, and decline; profits fall when sales stagnate and begin to drop.

Technology and Market Forces Driving Industry Life Cycles

Industries formed around fundamental new innovations will follow an S-shaped development. Industries such as cars, telecoms, and electric power have century-long industry life cycles and the IT industry is about 50 years into its life cycle.

In the earliest phase of the life cycle, technology and product development are the major drivers in industrial development. Early products are expensive and unreliable and the market is very small, typically consisting of business or military customers who can use the new technology to achieve breakthrough improvement in some functions of their own core business. Early steam engines for powering an entire factory in the nineteenth century are a typical example. Early computers used for code breaking during World War II are another. These early users are picking low-hanging fruits by using the new technology in niche areas where it can achieve radical productivity improvements. In this early phase, no dominant product architecture has been established and there are still complementary technologies and products missing for a market takeoff. Once a dominant design is in place, network effects will enforce this new standard and economies of scale will start to drive prices down. The new technology also needs a number of complementary technologies and supporting functions. In the early electrical industry, power plants and the electric grid had to be built to establish the market. Inventing the electric light bulb was a very focused research effort to develop a killer application that would drive market growth. This cluster of complementary industries needed to exploit a new technological innovation is called a development block.

Once a dominant design and a development block are formed, the new technology can begin to enter the mainstream market. This is the start of a long boom, where gradually falling prices will drive penetration and usage from the expensive high end into the mass market. During this phase the new industry is highly visible with high profits and growth attracting talent and capital. The industry is now shifting from being technology driven to being market driven. This phase is called the tornado or the construction cycle.

Finally, growth begins to slow down and price pressure forces the industry to focus on efficient processes and rationalization. In this phase, called the rationalization cycle, all other sectors of society learn how to fully exploit the potential of the new technology. Penetration of the new technology is massive and the impact on society is very large. The now mature industry is taken for granted and will be a platform for the next major life cycle.

As an alternative development, the industry itself gets a second wind from a new technology with the potential to propel it to new heights. In the telco industry, a number of technology shifts have ignited new growth in an industry that was about to exhaust its growth potential. Technology shifts in the telco industry have been from electromechanical switches to digital switches, from digital circuit switching to packet-based routers, from fixed to mobile telephony (Lind 2003).

Disruptive Innovations

A disruptive innovation is a new technology beating an old technology from below. That is, the new attacking technology is built on a simpler, hence cheaper, product architecture than the old technology. The attacking technology starts by finding new buyer groups who could not afford the old technology. The price is typically 10 times cheaper and the new technology quickly builds a large user base. As the market grows, the experience curve drives prices down much faster than for the old technology and the law of exponential growth in performance puts the new technology on a faster growth path than the old technology. The new technology begins taking market share from the old technology in the low-end market and the old technology gradually loses sales volume. As time passes, the growth path of the new technology reaches and crosses the slower growth path of the struggling old technology and the attacker wins.

Examples of disruptive innovations are the PC winning over the traditional mainframe and Unix computers, minimills for steel production winning over traditional integrated mills, and how the smaller 3.5 inch disk drives won the market over 5.25 inch disk drives (Christensen 1997; Christensen, Verlinden, and Westerman 2002).

Architectural Shifts in IT and Other Industries

Shifts in dominant architecture have been observed in a number of industries and technological systems, due to changing markets or new technologies. The new architecture is often less complex, more flexible, and cheaper than the old architecture.

In the transport system, canals were replaced by railroads, which were replaced by trucks and cars. The age of mechanical machine power started with very large steam engines that shrunk in size over time and were eventually replaced by combustion engines. Electrical engines are now gradually replacing the combustion engines (Grübler 1998). There are over 200 years and a number of shifts in dominant technological architecture between the first industrial steam engines and the electric toothbrush. The IT industry has gone through three major shifts in dominant IT architecture during its first half-century. First, mainframe computers were replaced by Unix-based minicomputers as the dominant IT architecture. Then the minis were replaced by PCs, and in the future we will see how the PC is attacked by sub-PC computers such as handhelds and embedded processors.

The emergence of dominant architectures in IT follows from the very strong network effects in software. Once a dominant design is established, the switching costs are extremely high, according to Metcalfe's law. Over time the complexity of the dominant software platform increases and it becomes overburdened by its own legacy. The software in mainframe computers is about 20 million lines of code, the same size as the traditional telco central office switching software and far less than the size of the Windows operating system. As software grows in size, complexity diseconomics is making it more and more unmanageable, with complexity increasing approximately as the square of the software size (Brook's law). At the same time, Moore's law is making less complex hardware platforms competitive and eventually the old software paradigm is swept away by the next wave.

The rise and fall of dominant architectures in the IT industry is therefore a consequence of Metcalfe's law (enforcing establishment), Moore's law (making the next cheaper hardware platforms good enough to compete), and Brook's law (ossifying old software architectures). Not all technology shifts strictly follow the theory of disruptive innovation, even though it is a valuable model for understanding technological change.

Empirical Support for Postmaterialistic Value Shift

The claims about value shifts in the population can be supported by empirical cross-cultural value studies undertaken in a number of countries during the past 30 years: RISC (1978–), Sociovision (1974–), European Commission (1970–), Inglehart (1990), and Hofstede (1980). The studies show that during adulthood most people do not change the values they have acquired by age 25. This fact can be used when analyzing the values of people born in the 1940s, 1930s, 1920s, etc., as a way of indirectly probing backwards to uncover value changes during the entire twentieth century.

These studies show values slowly changing as countries develop from traditional societies via modern industrial societies to postmaterialistic knowledge societies. For each decade, values change at a steady rate and in the same direction. This is an inexorable process as one generation is slowly replaced by the next. The studies show consistently how hierarchical and authoritarian values are eroded and replaced by egalitarian self-reliance, how unbridgeable inequality is replaced by equality, how collectivism is replaced by individualism, how masculine competitive values are replaced by feminine cooperative values, and how general trust in others increases. The ongoing value shifts also include a reduced need for rigidity, order, and structure. This goes with an increased acceptance of ambiguity and tolerance for unconventional lifestyles and other cultures. Materially secure, well-informed individuals with growing self-confidence are placing

Table 6.2   Long-term value shifts in society (various studies)

From	To
Power distance	Equality
Masculine	Feminine
Uncertainty avoidance	Security
Collectivism	Individualism
Traditional authority	Rational authority
Materialist	Postmaterialist
Limitations	Opportunities
Local	Global
Survival	Well-being
Seriousness	Pleasure
Structure	Flexibility
Hierarchies	Equality
Masculine competitiveness	Feminine cooperative

less importance on just living to make money and more value on self-fulfillment and well-being. The surveys show how the level of happiness and well-being in the population grows with rising GDP as societies move towards postmaterialism. Table 6.2 gives a summary of identified value shifts in the studies.

7

Technological Conclusions from Scenarios

Creation of four different scenarios gave us the opportunity to place our-selves in the middle of four distinctive worlds, and experience the simulated events as though we were living them. By analyzing the scenarios, what we really do is rehearse the future. In this way, a set of statements about wire-less development in 2015 can be identified by observing the surrounding environment. In this chapter we introduce these statements as technical implications of the scenarios and we formulate them in a positive manner. Assuming they are true in 2015 means that the underlying problems and bottlenecks we face today have been solved by then. We have divided them into three sections: (1) system technology, (2) mobile terminals, and (3) services. This is the first step towards identifying the key research challenges for industry and academia.

WE is in many ways the most positive of the four scenarios, describing a rapid development of technology and new types of service, as well as a dramatic increase in their use. Consequently, it is the most demanding scenario from a technological viewpoint, and many of the technical implications are therefore derived from this scenario.

System Technology in 2015

The system infrastructure will look quite different in 2015 compared to today. One reason is the need to build cheap and simple systems, another is that the large variety of services offered will drive the development towards a large variety of access networks.

The Wireless Infrastructure Will Be Heterogeneous

In all four scenarios, but less so in BM, wireless communication is provided by a large variety of systems. This is especially accentuated in WE and to some extent in RH, where only a few air interfaces and protocols are standardized in a traditional telecom manner. We see in our scenarios self-deployed broadband access points working in unlicensed spectrum, like WLANs, complementing the cellular systems at hotspots. Large capacity is also provided in cellular systems through picocell solutions.

DAB (digital audio broadcasting) and DVB-T (digital video broadcasting, terrestrial) provide spectrum-efficient broadcast services over large areas. Technologies like Bluetooth offer very short-range wireless access. Together with HAPs (high-altitude platforms), satellites, etc., all these systems make the future wireless world very diverse, complex, and difficult to integrate. Most of the time, this heterogeneous infrastructure is not hierarchically organized, meaning that different systems are connected directly to each other without centralized control from a higher system level. Distributed control solutions are chosen instead. Base stations and access points are adaptive, self-configuring, and use smart antennas to increase capacity and reduce interference where resources are scarce. In all scenarios the wired backbone network is packet switched and is available in most homes and offices, either directly (through wire or fiber) or through BWA (broadband wireless access).

Efficient and Very High Rate Air Interfaces Will Exist

Wideband hotspots and advanced broadband services, especially in WE, assume the existence of air interfaces capable of 10–100 Mbps for outdoor wide area coverage and up to 1 Gbps for very short-range personal communication. These data rates are required to support mostly short-range, cheap, and non-real-time services such as media downloading and fast synchronizations. Real-time connections, perhaps supporting advanced video-conferencing, can also be offered but at a higher cost.

Traffic Will Be IP Based and Networks Will Be Transparent

In all four scenarios, fixed and wireless traffic is packet switched, based on enhanced versions of IP (e.g., IPv6 and later versions). Networks are transparent for packets at the IP level, making it easy for service providers to develop and offer new and customized services to large numbers of users.

Much of the Access Infrastructure Will Be Ad Hoc Deployed

Reducing the cost for building wireless infrastructure is an important issue in all four scenarios, being necessary to decrease the cost per transmitted bit and boost traffic. Ad hoc deployed access points (by users or operators) will constitute a large part of the access infrastructure, especially in the case of broadband hotspots or other solutions offering mostly indoor coverage. A lot of traffic will go through spontaneous ad hoc created networks (e.g., users sharing information between two portable computers or a video camera sending a movie to a backup server). These kinds of system will mainly operate in unlicensed bands (e.g., 2.4, 5, 17, 60 GHz) in the scenarios, except in BM, where very little unlicensed spectrum is available. Ad hoc deployments will be present outdoors as well, but in much smaller proportions. The reason is that unlicensed operation and coexistence of different systems in the unlicensed spectrum still lead to inefficient spectrum utilization and make it impossible to protect the coverage area of an access point from interference created by neighboring access points, which might be competitors or even radio jammers deliberately obstructing traffic. In some cases this issue is solved by agreements between different owners of access points regarding the operation type, priority, coverage radius, etc. In other cases the situation might be very bad, with a completely chaotic environment where everybody tries to use the spectrum but in an uncoordinated way.

Cost per Transmitted Bit Will Be Very Small

The very rapid growth of wireless services and applications, like in WE, is based on very low transmission costs. The cost per transmitted bit in 2015 is about 100 times smaller than today. This is possible due to the proliferation of new types of systems that are cheap to plan and install, easy to operate and maintain, and running best-effort services (e.g., self-deployed systems). These systems typically provide high capacity in small areas, hotspots. Cellular systems provide coverage over large geographical areas.

No Harmful Radiation from Base Stations

Due to the possibly real and perceived risks concerning electromagnetic radiation from wireless networks, especially in SM and RH, all base stations and other access points are designed to meet very low limits on radiation levels. In urban areas these solutions involve low-power, wideband, and short-range communication inside pico- and microcells instead of the old macrocellular approach.

Decreased Power Consumption in the Wireless Systems

The increasing need for power to drive computers, routers, server parks, battery-powered devices, etc., is an environmental problem on a global scale. The strong environmental concern among the public, especially in RH and SM, has pushed the industry to radically decrease power consumption in wired and wireless communication systems.

Mobile Terminals in 2015

Terminals will come in many shapes and forms in 2015. Some will be cheap, simple, and very basic; others will be advanced and expensive. Demands on flexibility, adaptability, user friendliness, usage time, and radiation levels are driving the development.

Terminals Will Have a Wide Range of Shapes and Capabilities

In all scenarios a range of different terminals will be offered to end consumers and other users. The terminal of 2015 is a common device in most people's pocket and its capabilities range from basic to very advanced functionality. Industry markets everything from very simple and service-oriented terminals such as targeted at Elders in RH, to the most advanced 3D-capable screen-enabled device for handling the very high bandwidth requirements of Mobile Professionals in WE. Moklofs are able to customize the design of the device itself and its interface, whereas Yupplots focus more on very secure services that save them time when making payments, banking, etc.

Wireless Terminals Will Be Cheap, Very Small, and Modularized

Most of the terminals are modularized with clearly defined interfaces between modules, for example radio unit, input/output unit, screen, and keyboard. The different modules are also sold as separate components and are then used to create individual device systems, especially in RH and WE. Many of these cheap and service-dedicated wireless components are integrated into fashion items like clothes or accessories or into portable computers and PDAs. Such wearable device systems are managed as BANs (body area networks). Routing of data through many devices or modules is simple, easy to implement, and easy to set up and run for the end user. System-on-chip technology, where the processor and the wireless communication unit are integrated into a single chip, has enabled such ultra-small and low-cost components.

Usage Time Without Charging Batteries Will Be Very Long

In WE the wireless terminals have a usage time of at least one week without charging, even if heavily used. Users in all scenarios demand long usage time, in standby mode and while using the terminal. Power management in the terminal is a key solution but not the only one. Battery capacity has also improved and is two to three times larger compared with 2000. Portable power sources like fuel cells provide much larger capacity, but they can also be dangerous if used improperly, so their use is still somewhat restricted.

User Interfaces Will Be Highly Developed and Advanced

Most terminals feature several functions that greatly facilitate communication and use. Advanced solutions are common for voice control, touch screens, and interactive control. Even traditionally slow adopters, such as Elders, use advanced wireless services, especially in WE and RH. The same service can be provided on different input and output devices, such as a big screen on a wall, a desktop or laptop computer, and a PDA for people on the move. Advanced display technology allows the virtual size of the display to be much larger than the physical size of the screen. Self-learning devices help the user to personalize the interface and to filter and organize information coming from various service providers, content providers, or the internet. Advanced voice interfaces can eliminate buttons, so the size of the terminal can be very small. In summary, an advanced and intelligent user interface is crucially important and a most powerful marketing tool for terminal vendors.

M2M Will Be Everywhere

Machine-to-machine (M2M) communication is enabled by the integration of very cheap communication modules into portable computers, digital cameras, house appliances, industrial devices, vehicles, etc. In the WE scenario, M2M generates a large amount of traffic. PANs and BANs, together with security and logging systems, are easy to implement. The amount of data exchanged between different devices will range from a few kilobits sent by a remote control, to a few gigabits when transferring a movie from a digital video camera to a TV set. M2M communication is implemented by $1 radio modules, similar to Bluetooth. All personal computers, PDAs, digital cameras, and even some home appliances (fridge, TV set, etc.) have a module like this preinstalled.

Wireless Devices Will Be Harmless to People and the Environment

One of the major obstacles causing a slow development of wireless technologies and services in SM is real and perceived health problems due to electromagnetic radiation from terminals. Several solutions to reduce radiation and absorption of electromagnetic energy by the human body exist. Smart terminal antennas can direct the radio waves from the user's body or head. The transmitted power of the terminal is reduced to very low levels, but this has a negative impact on infrastructure design, mainly because of the high density of the access points or base stations required.

Mobile Services in 2015

The variety of services will be very large in 2015. Some will be simple, whereas others will be very advanced. Some services will be generic and offered to users all over the world, independent of location, whereas others will be personalized, location based, and context dependent. Important issues include making the services independent of the infrastructure, offering seamless roaming between systems, and providing security in transmissions and use.

Wireless Services Will Become a Commodity

In WE and to some extent RH, access to wireless communication is provided in a transparent and seamless way to the users. Almost everyone owns at least one wireless terminal, often several, and the services are adapted to most user demands. Cellular and other types of centrally controlled infrastructure offer cheap and basic services to everyone. User-deployed access points are available at a very low cost, which means there is wireless access (with high bit rates) inside or in the proximity of almost every house. Always having wireless access in every corner of your home is as natural as always having a power outlet nearby. The basic service types are largely generic, standardized, and cheap; they are offered almost everywhere. Location-based services (e.g., positioning) and automatic user identification are very common.

Services Will Be Independent of Infrastructure and Terminals

Most services can be provided over different networks, on different types of terminal, using different user interfaces. The user will be able to choose the receiving device and the service format according to preferred screen size or price, for example.

Telepresence and Emotional Communication Will Be Available

Very complex services are available, placing very high demands on networks and terminals. Examples include wireless services involving real-time face-to-face communication complemented by fast sensorial feedback and 3D screens.

An almost real, but still virtual, sense of presence can be achieved between people communicating—telepresence. This enables advanced family in-tranets, where all members of the family can easily transmit their feelings (love, compassion, aggression, fear, etc.) over the communication channel. Emotional communication is especially important in RH, where frequent global traveling (for work or pleasure) splits families. Another example of an important and demanding service is a virtual meeting. Moving towards a more nomadic, or meeting-oriented society (Dahlbom 1999, 2000), this service is increasingly important and popular. Real-time multicasting facilitates high-quality group meetings. The data stream is characterized by real-time full-motion video with very rich sound capabilities.

Content Will Be Personalized According to User Demand and Location

Services based on identification and positioning of the user are common in all four scenarios. Content becomes more and more personalized and context dependent. Many services are provided in an instant turnkey fashion and they disappear when the user moves to a different location. Virtual information agents, portals, and information brokers assist the users in finding, filtering, and personalizing information. Two examples are personal assistants that keep track of a user's meetings, tasks, alarms, or address books; and virtual guides for tourists in unfamiliar cities or museums.

Global Roaming and Seamless Services Will Be Possible

In all scenarios users expect global roaming for at least the basic and simple services, like voice and text messaging. With the exception of SM, most users expect the same type of seamless coverage when they travel across the world, using more advanced services as well. All different technologies, air interface standards, communication protocols, etc., go unnoticed by the user. The terminals are automatically configured and no settings have to be changed when the user moves into a new cellular system or another type of network.

Broadband Services Will Be Available for All Transportation Systems

Wireless broadband connections are available in most vehicles, such as cars, trains, boats, and aircraft. User mobility is high in all scenarios. Many users travel longer distances in cars and on public transport. The user will be able to browse the internet, connect to their company intranet, or use other services in a very reliable fashion, even while moving at very high speeds (e.g., in high-speed trains, regular jet aircraft, and supersonic aircraft).

The End User Will Be Always Best Connected

Wireless services will be provided over a multitude of network types, for example cellular systems, DAB, DVB-T, WLAN, Bluetooth, HAP, and satellites. The concept of being always connected is very important in all scenarios, and solutions include flexible terminals that are able to handle different standards of air interfaces and data protocols as well as storing large amounts of data when passing a hotspot, for offline use. The system or the terminal simply chooses to access the best network by considering bandwidth demand, time, cost, or any other quality-of-service parameter. The user is always best connected even though the most efficient way to access a wireless service might vary between different locations or over time. Multirate and multiresolution services allow a connection to be maintained even if available resources (power or bandwidth) are low. Multicasting is employed, especially for certain types of services, e.g., multiparty video-conferencing or media distribution.

Powerful Computers Will Be Everywhere

Computers are moved into and hidden in our immediate surroundings (e.g., walls, clothes, furniture, and appliances). This is especially accentuated in WE. Several processors share the most demanding tasks, making even small computers very fast and powerful. Users access these resources by wireless communication through their terminals. Powerful computers are everywhere, but unlike today's bulky desktop or laptop computers, they are invisible to the user. They can be easily accessed and many of the heavy computational operations are performed outside the mobile terminal. Some terminals can therefore be thin and cheap, but they will be very dependent on the infrastructure. Advanced sensor technology is used to build smart spaces, environments that can interact with wireless terminals and adapt to a wide range of user needs. As an example, voice commands are used to customize these spaces, for example preparing a room for a business meeting with an agenda and the participants' slide presentations or starting a relaxation program in the same room (based on sounds, light, and smell sensor data) after a hard working day.

Very High Levels of Security Will Be Provided

Protection of wireless transmissions against interception, viruses, and jamming is an important issue in all scenarios. Guaranteeing security, integrity, and privacy of services is very important for most users, particularly when it comes to financial transactions such as mobile banking and payments, or access to corporate intranets. In BM especially, large efforts are made to protect the privacy and integrity of the users. Measures to combat unauthorized tracking of users and their behavior are continually being developed. Even though advanced services that are based on information regarding positioning, automatic identification, or wireless banking, etc., are common, it is almost impossible for someone without authorization to get access to private information about the user (position, last shopping places, etc.). Privacy is guaranteed since personal or other sensitive information is secured and made public only on request from authorities or the user.

Part III

Challenges for the Future

8

Challenges for Technical Research

Our approach is to create broad scenarios shaped by "mega-trends" within technology, society, industry, and among users, and then use them to identify important research areas. We believe it is a fruitful approach, but we do not believe we have identified all, or even most, important fields where a lot of work has to be done. The scenarios represent our visions of the future and are open to further interpretations. Hopefully, the reader of our scenarios can draw other conclusions about technical implications, thus identifying other important research topics. Many of the areas identified here are well in line with other future-oriented studies, for example WWRF (2001) and Bria et al. (2001), and are well supported by the scenario descriptions.

The research issues we feel are the most important for creating a positive wireless future are (1) low-cost infrastructure and services, (2) seamless mobility, (3) new and advanced services, (4) usability and human–machine interface, and (5) health and environment.

Low-Cost Infrastructure and Services

The cellular systems of today are primarily designed to provide cost-effective wide area coverage for a rather limited number of simultaneous users with moderate bandwidth demands. In order for new wireless services (e.g., multimedia services) to succeed on a large scale, they have to be widely available, simple to purchase and access, and affordable to large numbers of users. This is not only a question of high bandwidth and quality of service. The cost for the user must be comparable for services offered in the current generation of cellular systems.

One problem with conventional cellular systems (2G, 3G, etc.) is that they don't scale in bandwidth in the economic sense. A large part of the cost for building the infrastructure is related to network planning and site work. Economies of scale, and certainly Moore's law, are not applicable on site acquisition, roadworks, erecting towers, etc., in the sense that they do not follow the same price/performance evolution like electronic components. Also, the cost of the wireless networks depends rather weakly on the basic radio technology (e.g., the air interface) since current modulation and signal processing technologies are quite advanced and very close to the theoretical limits (Shannon's law) that not even a radical improvement in processing capabilities will significantly improve performance. If we maintain the area of coverage, capacity in terms of the number of users, and quality of service, the infrastructure cost will be directly proportional to the user data rate (Zander 1997). In other words, the higher the data rate a service requires, the more it will cost to deliver. In the end this will translate into a higher price for the user, meaning that advanced multimedia services will be expensive. Today's users are accustomed to being connected anytime and anywhere (i.e. large coverage areas and high availability), so these parameters can hardly be compromised. If affordable multimedia services are to be possible, i.e. higher data rates at constant or lower cost per user, either some of the other QoS parameters have to be sacrificed or architectures with radically lower cost factors have to be developed.

There are two ways of approaching these cost limitations of broadband wireless systems. One way is to go, from the outset, for a larger number of small, cheap, and ad hoc deployed access points, similar to WLANs. That means compromising anytime/anywhere coverage or giving up on quality of service, but it will make wireless broadband communication available at least at hotspots. The second option, more suited to established cellular operators, is to slowly improve the capacity of their networks, thereby increasing the number of supported users by upgrading to more expensive, but more advanced base stations and smart antennas.

Due to the high cost of developing, building, maintaining, and operating wireless infrastructure in general and cellular infrastructure in particular, issues relating to efficient resource management will become increasingly important. By "resources" we mean spectrum, power, and available infrastructures. The large number of services and the multitude of service types expected in the future are the main challenges for resource management mechanisms. The necessary data rates, for example, may vary substantially, from perhaps 10 kbps for a certain user under certain conditions, to 100 Mbps for another user under more demanding conditions. We will here discuss a few issues that can lead to cost reductions in the future wireless systems, especially in the case of cellular operators.

Even today, techniques like RoF (radio-over-fiber) allow new network configurations and protocols, leading to cost savings compared to traditional cellular configuration if the optical fiber is already available or is cheap to deploy. High-altitude platforms (HAPs) or low earth orbit satellites (LEOs) may provide coverage and high capacity, sometimes in a cheaper manner than ground infrastructure. These techniques may well be helpful for future 3G systems in a long-term perspective. A promising development is the use of smart antennas to improve the link quality and achieve higher capacity in the networks. Multiple-input multiple-output (MIMO) channels are a good example, offering great potential, at least in some environments. The air interface protocols of the future have to provide several times better resource utilization and spectrum efficiency than today.

Better utilization of resources and lower costs can be achieved if several operators share the same infrastructure (infrastructure sharing) or the same spectrum (spectrum sharing and cofarming). In the case of infrastructure, both radio access network (RAN) sharing and transport network (TN) sharing should be considered. One example is implementing the connections between the base stations of cellular systems with the same cheap and shared Ethernet connections as for WLAN. In the future, the optical fiber infrastructure will provide immense bit capacity, and sharing it among different service carriers will be the key to low cost. In the case of spectrum, we will need efficient algorithms for avoiding collisions between transmissions from different systems coexisting in the same spectrum.

Since the demand for different services typically varies over time, the available spectrum resources can be used more efficiently by advanced re-allocation procedures (dynamic spectrum allocation). Peer-to-peer communication demands high capacity during working hours (from 9 A.M. to 5 P.M.), when spectrum could perhaps be borrowed from broadcasting systems. During evenings, when people consume more television, less capacity can be allocated for telecommunication and more for broadcasting. During late-night hours, the spectrum can be used for broadcasting or multicasting information and media content to caching-enabled terminals. The integration of personal communication and broadcasting systems is a hot issue for the future.

It is a considerable challenge to provide high capacity wherever it is needed, instantly and temporarily. Examples of this could be for an event company to provide voice, messaging services, and web surfing to a large audience at a music festival or for a news company to broadcast images of a crashed airliner in a thinly populated area. In both cases, consumers or media companies might be willing to pay for this type of temporary broadband service. Research has to be done on techniques for dynamic allocation of resources in time and space in a very flexible manner.

A consumer of wireless services typically retrieves more information from the network than he or she sends, making the traffic pattern asymmetric. This could lead to cheaper infrastructures and terminals, as well as longer usage time before recharging the terminal battery, and increased spectrum efficiency. How to make better use of this traffic asymmetry is an interesting topic.

Ad hoc deployment of the access points and minimal work on planning and maintenance are the strengths of today's WLANs. Technologies like WLAN and Bluetooth will be embedded in all future communication devices and personal computers. This will require the development of policies for routing signals through many devices organized in ad hoc networks (e.g., M2M communication and multihopping). Decentralized resource management will be an option when network topologies become large. In a very diverse and heterogeneous environment, complexity means that a decentralized control scheme is preferred to a centralized one. Future research should focus on decentralized resource allocation, possibly involving software agents able to trade resources in real time according to supply and demand.

The increasing demand for new wireless services will lead to the deployment of new wireless systems of different types: new cellular systems, wireless LANs, satellite communication systems, etc. In a few years there will be a need for spectrum reallocation. More available spectrum for an operator makes the network cheaper. It is very important to identify and release new frequency bands for personal communication, for licensed and unlicensed operations.

Seamless Mobility

As the wireless world becomes increasingly heterogeneous, with many types of access network, a potential problem is the difficulty of controlling and managing this complex environment. When the complexity increases, it becomes more and more difficult to design centralized systems for controlling the networks, such as systems providing seamless roaming and billing solutions. Moreover, as the complexity increases, the modular design of the cellular networks becomes increasingly wasteful of bandwidth and energy. At the same time, it is very hard to predict and manage complex interactions in decentralized (distributed) systems as well, making this a difficult problem.

One reason for the importance of complexity management is that most users have come to expect a rather high quality of service, for example seamless roaming between access points within a single network and between networks operated by different operators. When the number of systems operated by different actors increases, this problem will be accentuated. This is not only a technical problem, but also an administrative and business problem. From the user's perspective, the important thing is the services offered and their quality, not how the systems look or who operates them.

If seamless services, always best connected, and infrastructure-independent services are to be possible in the future, different systems and access networks have to be integrated into a "single" network, transparent for data traffic. There are several ways to achieve this integration. One way involves standardization of interfaces at different system levels. It is therefore necessary to develop networks and terminals that can switch between different air interfaces and protocols. Multimode and adaptive radios will enable this feature by employing software implementations on a common hardware platform or system-on-a-chip technologies. Research should be carried out on how to implement low-cost and efficient multimode and adaptive radios for terminals and infrastructure.

User access to information will be provided in real time and non-real time. If there is no delay constraint, then the non-real-time choice is cheaper to deliver, but still well suited to a large number of services based on downloading and caching data for later use (multimedia messaging, news, music, movies, etc.). All this traffic can happen in off-peak hours. The use of information gathered in this fashion can indirectly influence performance in a mobile situation. Research should be carried out into the implementation of services based on collecting, memorizing, and organizing information in the terminal in a smart and personalized manner.

Due to the wide range of terminal capabilities, seamless services need to be provided in a scalable manner. The same application should be adapted for low and high data rates, for small and large screens, and for low and high price ranges. As the number of customers increases, the resources have to be scaled up to the new requirements in a linear manner. Research on scalability of services as well as techniques for supporting different QoS levels will be very important in enabling viable business models in the future wireless systems. Advanced signal processing in the areas of voice and image coding has a large potential to contribute. Multidescription coding, for example, divides video data into equally important streams so that the decoding quality using any subset of them becomes acceptable. Higher quality is obtained by increasing the number of descriptions, so the probability of losing all the descriptions is small. There are already a few implementations that look like very promising ways to provide scalable services. More research is needed on this kind of signal processing.

Future consumers of wireless communication will be increasingly mobile, moving from place to place to meet friends, do business, watch movies, etc. As the requirements on bandwidth and infrastructure density grow, mobility management could become a problem. User identification, handover procedures (when moving from one cell to another or from one system to another), roaming, billing, etc., will consume an increasing amount of resources and time, contributing to the perceived delay of the services. Research is required on how to efficiently handle these aspects of mobility.

New and Advanced Services

In the future a multitude of new service types will be introduced. In several of these services the typical content will be location dependent, for example enabling services to provide you with a city map when you arrive at an airport, giving special offers when you pass a shop, or providing online games when you wait at a railway station. Some content varies from place to place or environment to environment, and the personal terminal is able to memorize and collect information while moving in the network. Future investigations are needed into developing context- and location-aware services plus investigations into defining the requirements for the terminals and the infrastructure.

Multiple-party meetings in real time are an important and demanding future service, which can be further extended to so-called telepresence. Telepresence is a virtual meeting that provides the illusion of actually being somewhere else, an illusion very close to the real thing. The bandwidth required for telepresence is expected to be about 100 Mbps (with efficient data compression techniques and fast sensory feedback). Techniques are needed to enhance telepresence over future wireless channels and sensor technology is needed to enable it. Multicasting can be used to increase spectrum efficiency when offering multiparty telepresence.

Moving computers into the network and making them invisible to the end user are important features of the future wireless world. Being always connected and having access to computational and communication resources will lead to ubiquitous and seamless services, enhanced by smart spaces for example, with a multitude of displays and other resources surrounding the user. This will require the development of suitable services with advanced user interfaces having appropriate sensors, good and cheap output mechanisms, and smart devices.

Security will be one of the most important service features in the future. Better techniques to ensure protection of data and user privacy (position, traffic patterns, private information, etc.) have to be developed. Ways to provide very secure connections and secure data transfers, at least in services like electronic payments or banking, have to be studied. Unauthorized tracking of users or access to personal and private information has to be avoided. The whole area of security, privacy, and integrity has to be approached seriously.

With the increasing use of new and demanding services (e.g., multimedia services) there will be a need for efficient and very high rate air interfaces. Air interfaces of at least 100 Mbps for wide area coverage and up to 1 Gbps for very short-range personal communication seem to be needed. These data rates are required to support services like media downloading, fast synchronization, and advanced videoconferencing. OFDM (orthogonal frequency division multiplexing) and UWB (ultra wide band) are two examples of candidate technologies.

Usability and Human–Machine Interface

Usability refers to how easy it is to handle a product or service. Ways to improve usability include shortening the time to accomplish tasks, reducing the number of mistakes made, reducing learning time, and improving people's satisfaction with a system.

The user interfaces of wireless terminals are currently based mainly on physical or virtual buttons (existing on a small touch-sensitive screen), input devices like microjoysticks or roller wheels, and even simple voice recognition. The output capabilities of wireless devices are also hampered by the small size of the devices, battery constraints, and the cost of displays and sound cards. This has important limitations for usability, especially when we consider senior citizens, disabled persons, and other people with special needs or requirements. The user interface of a physically small device has to have a small scale, at least today. Furthermore, the user interfaces of wireless devices follow proprietary models and are neither standardized nor very intuitive for the user. If the user has learned to be efficient and enjoys the interface on one device from a certain vendor, they will be very frustrated when having to learn the interface on a device from another manufacturer.

Research needs to focus on ways to handle these problems, for instance through splitting the single wireless device into simpler information appliances, each taking care of a specific task, thus providing better opportunities for user interfaces that are easy to understand and use. If the radio part of a mobile phone were embedded in a device of its own, no direct manipulation would be needed. This micro base station could be placed anywhere on a user's body or in their vicinity. A specialized device for user input could be connected. This device could have a larger display than battery constraints would allow on an integrated device. In addition, control through new modalities, such as voice, could be used to control this personal communication system, thus completely removing the need for buttons on a device or virtual buttons on a display. Augmented reality using goggles and fold-out screens are promising topics that deserve investigation.

For users with special needs, such as people with failing eyesight, hand tremors, or rheumatism (all of which make it hard to manipulate small objects), certain parts of a personal network could be exchanged for specialized devices better suited to these users' needs. Networks like these could also interact with fixed devices surrounding the user. A user could seamlessly move information from their personal display to a larger display with high resolution somewhere in their vicinity.

Research should therefore be focused on developing a more user- and human-centered system of appliances where the technology of the computer disappears behind the scenes into task-specific devices (Norman 1998), perhaps integrated into everyday things. But fragmentation of the user interface technology does not solve the whole problem. Improved design is still required to ensure uniform and intuitive interfaces.

Health and Environment

Use of communication systems, especially wireless systems, raises several important issues of health and environmental impact.

The effect of electromagnetic radiation on the human body is an area of crucial importance for the whole wireless community. As yet, no generally accepted scientific research has proved that use of wireless terminals is dangerous, at least not with the radiation levels allowed today. However, there is a need for more research in this area, especially studies to examine the long-term effects of radiation on the human body. Current standards and limitations on mobile phone transmitted power are based only on the thermal effects of electromagnetic radiation. There are some indications that there might also be radiation effects on DNA structure and cell reproduction. The protection standards and maximum limits will probably have to be modified according to all possible negative effects of electromagnetic radiation.

The impact of communication infrastructure on the environment is another important issue. One important area is the power consumption in the systems to drive computers, servers, base stations, etc. The environmental impact of producing the required electric power is not negligible. Towers, underground cables, high-power transmitters, and access roads to base station sites can also have a negative impact on the environment and on people's esthetic values. The likely development is towards an increased density of ground infrastructure for communication, which means future systems will have to be integrated into the environment in a sustainable way. Another way to tackle this issue is to move the infrastructure into the sky by using HAPs (high-altitude platforms) or different types of satellite.

Increased use of wireless services also has a big potential to affect the environment in a positive way. New and advanced services might decrease the growth in travel. The transportation sector might benefit by being able to rationalize the distribution channels (fleet management), saving fuel and decreasing congestion. There are several other benefits too.

In conclusion the impact on health and environment is a cross-disciplinary area that needs to be investigated by examining radio communications technology, medicine, biology, environmental engineering, and business.

A Need for Cross-Disciplinary Research

One important conclusion from this work is that the academic research tradition with well-established and very specialized areas of research will not suffice in the future. The need for parallel cross-disciplinary research is becoming increasingly important.

The reason is that research needs to widen its focus beyond creating new technology and knowledge within specific areas, otherwise we run the risk of developing new technologies with little or no connection to the needs of users or the industry. Problems outside the computer simulation program or the laboratory very rarely lend themselves to formulation as well-defined entities in the same way that academic research groups are organized. The world *is* cross-disciplinary. It is crucial that academic research also takes real-life problems as a starting point, not only problems from the academic tradition. Instead of focusing on developing a new high-bandwidth air interface, for example, research can begin by asking whether this has the potential to solve a problem for users or the industry. What if increased bandwidth is not a solution to a relevant problem? It might be that wireless systems are too expensive and research should focus on developing extremely cheap base stations which can be deployed very easily. Both research approaches have their merits and both are needed for us to create a good future.

9

Challenges for the Wireless Industry

Introduction

There are several important challenges facing the wireless industry in the next 10–15 years. This chapter highlights topics we consider especially important. The views in this chapter are based on our scenario work and input we received from external experts. Our proposals are based on topics we believe to be critical for positive development of the industry and indicate how the industry might stumble if things go wrong or are left unresolved.

The Challenges

Threat from Disruptive Market Change

At first sight, the traditional mobile industry looks very impressive with advanced R&D, high revenues, and over 1 billion users. Products from the equipment vendors have a reliability, complexity, and sophistication unheard of anywhere else. But the weaknesses are there, just below the surface. Equipment and systems are complex and hard to control centrally. Complexity and small production volumes make the products very expensive and product development is rather slow. This is less of a problem when the only customers are large operators. But the telco vendors and operators, living in a world of long planning cycles and billion-dollar orders, are seriously threatened by attackers with a completely different business model—the datacom industry.

The hegemony of operators and their telco equipment vendors is built on the ability to own the users by locking them into a closed system and on controlling the only available spectrum. The traditional business models are based on centralized control and classically planned telco infrastructures with tower masts, cells, billing by the minute, handover, and roaming. It is not at all certain this situation will prevail in the long run. What the telco industry might overlook are new technologies that can be used as entry points to attack this status quo: IP, unlicensed spectrum, self-deployed networks, ad hoc and peer-to-peer networks, self-configurable network elements, and open APIs (application program interfaces). The players in the fast-moving IT and datacom industries are masters at exploiting weaknesses in incumbent business models and finding soft points for attack. They understand that users prefer cheap products here and now, if those products can meet their immediate needs for an acceptable quality of service. The market accepts unreliable and simple products if the price is low enough. This is a classical setting for a so-called disruptive market change (Christensen 1997).

Speed up the Process of Spectrum Release

Radio spectrum shortage is one of the most important inhibitors to further industrial development. If governments allocate abundant spectrum to the wireless industry, growth will be much faster and prices for the consumer will be lower. If governments are slow to release new spectrum, growth will be slowed. The process of global spectrum allocation is today handled by the diplomatic World Radio Conference (WRC), meeting every two or three years with the agenda being set between two and five years in advance. With the present slow-moving WRC process, it will not be possible to release significant amounts of new spectrum until somewhere around 2013. Therefore spectrum management needs new forms and the issue must be put high on the political agenda in order to speed up the process.

For historical reasons, most spectrum is locked in by legacy users, the military, and television broadcasters. In Europe the military controls on average 30% of all spectrum (Lightman 2002). Wireless consumer communication has been given less than one-tenth of all usable spectrum between 0.5 GHz and 5 GHz. Coupled with the rapid increase in wireless users during the past decade, this has caused a shortage of spectrum in many densely populated areas. This shortage is forcing operators to build unnecessarily expensive infrastructure. Growing usage will only aggravate this problem.

Apart from allocating more spectrum, the shortage can partly be alleviated in several other ways. Spectral efficiency can still be improved somewhat by using more computing power. Smart antenna technology can be used to focus the radio signal in a beam, and resource management can be developed for smarter sharing of spectrum between different user types.

3G and the Telco Debt Threat

An obvious threat to the wireless industry is the enormous debt left behind after the financial hype a few years back, in particular from the 3G auctions. In addition, operators are now facing future investments of the same magnitude for building the 3G networks. It seems clear this means substantial delays in rollout of 3G, putting a severe strain on the vendor industry as well. If the economic recession deepens, credit will be squeezed and the financial actors will then probably refuse to extend credit to industry. If 3G is not to become the technology that sent the industry into bankruptcy, the problems must be put on the table and dealt with. The unrealistic plans for several completely new physical infrastructures built within a few years must be brought down to earth. European governments, in particular, should alleviate demands on the 3G operators and allow unlimited infrastructure sharing and a slower rollout. It is better to have one 3G network in operation than five bankrupt operators. The business case for 3G would be more reasonable if it were allowed to grow organically with usage and if it were possible to avoid the large costs for erecting mast towers all over Europe when building new networks.

Complexity Management

In a future world with users seamlessly connected over a number of heterogeneous networks, complexity will be much higher than today. It will be unmanageably complex to have central control for billions of users operating various terminals to access many networks with handover, roaming, context-sensitive user profiles, billing, uninterrupted sessions, etc. The industry needs to address these issues and develop decentralized solutions for handling complexity. This calls for open APIs such as IP and adherence to open standard interfaces.

Radiation a Problem, Real or Perceived

The complicated problem of electromagnetic radiation from wireless terminals and base stations has to be taken very seriously by the industry. Even if, as many experts argue, the radiation levels permitted today are in fact harmless, they are a threat that needs to be dealt with. The problem is that no proof of danger is not the same thing as proof of harmlessness. If users are afraid of using wireless technology, it is a problem, justified or otherwise. The industry should consider taking precautions and should tell users how to minimize the radiation penetrating their body. It should develop modularized terminals where the radio unit is removed from the absolute vicinity of the most sensitive areas of the body.

Better Batteries in Wireless Devices

With the very rapid development of processor power (following Moore's law) and memory capacity, making new services possible, the power consumption of wireless terminals will increase dramatically. At the same time, battery capacity develops much slower. From a user's viewpoint, long usage time between charging the device is a very important feature. Most users are accustomed to quite long usage and standby times on their GSM phones. It will therefore be very hard to convince the mass market to use power-hungry services if battery time drops too much.

Usability and the User in Focus

In contrast to the technology-driven development of the past, the wireless future will become much more user driven. It is very important that the industry begins to focus on developing technology and services which solve real user problems. Usability and intuitive user interfaces are very important when access is through a tiny display. This is an area where operators, terminal vendors, and application developers need to synchronize their efforts. The failure of Wap, marketed as wireless internet surfing, is an example of a vendor-driven technology with no clear demand, at least not with the terminals, services, and networks available at the time. Usability is probably the single most overlooked area by today's wireless industry.

Cheaper Infrastructure and Viable Business Models

The current mobile cellular infrastructure has been deployed under a high-cost business model, maintained by high revenues from the users. The systems of today have been built under the assumption of spectrum shortage, reusable cells, uniform geographical coverage by one technology, a limited number of simultaneous users, moderate bandwidth demands, and a dedicated physical infrastructure with masts and cabling. This is not a viable way forward. Users will not be prepared to see their average wireless bills increase by several hundred percent, which is necessary if future wireless multimedia are to be carried over traditional networks. Therefore we need to develop innovative ways of providing wireless bandwidth at affordable costs in a world of many heterogeneous networks.

A number of issues concerning viable business models and technological solutions are relevant here. An example of an important cost driver to avoid in the future is the traditional macrocell infrastructure with tower masts. In urban areas this can be achieved by optical wireless, ad hoc networks, etc., or self-deployed cheap base stations simply connected to the fixed net. Another important issue is to create business models where market players can make money by benefiting from radically cheaper technologies and where all actors in the wireless universe can thrive.

A Phone for Everyone

There are less than 2 billion people in the world today that have a fixed or mobile phone subscription. Even though this is an impressive number, it also means there is potentially a similar-sized market of people who don't yet have access to a phone. This simple fact suggests there are two different strategies for the large telecom vendors. One strategy is to continue to focus on the rich industrialized world and the big NICs with large market potential, introducing complex and expensive technology plus advanced services. The other strategy is to focus on the other 2 billion potential users and develop very cheap but reliable solutions to satisfy basic communication needs.

All Industries Mature

Looking back in history, it seems evident that all industries, even though considered hi-tech in the early days, eventually matured and entered a phase with slower technological development, a phase with profitability being driven by efficiency in manufacturing and large volumes, leading to low production and distribution costs. There is no reason to believe that wireless data and telecommunication are an exception. For a few decades now, wireless communication has been a typical hi-tech industry with very fast technological development, high profit margins for successful companies, and almost exponential growth. This industry has also been, and still is, very much in the public eye. The question is not if, but when it will mature. Perhaps we are seeing the first signs today, for example a long-term trend towards lower profit margins and industry restructuring (acquisitions, mergers, and cooperation between actors, etc.). What if it turns out that there are no fundamental user needs other than voice and simple messaging services accessible anytime and anywhere? This need can be easily satisfied with today's technology. On the other hand, what we are seeing now might be nothing but a temporary dip for an industry that still has decades to go before reaching maturity.

10

Challenges for Key Regions

Each major world region has its own set of wireless strengths and wireless weaknesses. Even though the hi-tech industries are global in nature, these different capabilities will have a significant impact on the wireless industry and markets in the regions. The US is global leader in the IT and datacom industry but is lagging behind in mobile coverage and penetration. Europe is the global leader in GSM but risks losing this position during its current telecom crisis. Japan with iMode and FOMA is global leader in wireless data but the small market and the severe mishandling of the Japanese economy leaves many questions for the future. China is the largest mobile market in the world with great potential for the future but the Chinese vendor industry is still far from competitive. Korea and other Asian countries are also important markets. Countries such as India, Brazil, and Russia are slowly moving forward and they will be important consumer markets due to their size. However, we don't believe these emerging markets will have a signific- ant impact on the international arena or produce global players for at least 15 years. Therefore this chapter will focus on the US, Europe, China, Japan, and Korea.

US

The US is one of the single most important markets for IT and telecommunications products and services. For wireless technology and products, only China can be considered of equal or greater importance today and in the future. The sheer size of the population, almost 290 million people, high technological maturity, and a world-leading IT industry are important reasons for this.

In 2001 there were almost 140 million wireless subscribers corresponding to a penetration rate of about 48%, high on a global scale but significantly lower than the average 74% penetration rate in Europe (Swedish Trade Council 2002). Growth rates are high. In 1996 there were only about 44 million wireless subscribers and forecasts predict that the number will increase to about 245 million in 2006 (Swedish Trade Council 2001). Even though the US IT and telecommunications sectors are experiencing difficulties, just like in the rest of the world, the wireless sector is growing quickly, not only measured by increase in number of users but also by industry revenue, cellular user minutes per month, and number of SMS messages. Competition among cellular operators is fierce and, during past years, users have enjoyed significant price drops on tariffs for voice and simple data services.

An Immature Market for Mobile Services Waiting to Catch Up

Despite its very strong IT industry, the US lags Europe and Japan by two to three years when it comes to wireless services. The market situation in the US differs substantially from Europe and Japan. Digital voice services took off later and the US wireless data market is in its infancy but is catching up fast. There are approximately 5 million subscribers for wireless business data services and by 2006 that number is expected to increase to about 40 million (Swedish Trade Council 2002). SMS use is beginning to take off, with 1 billion messages sent in June 2002, up from 30 000 in June 2001. This is an impressive growth but from a very low level. In Europe 1 billion SMS messages are sent every day. There are two main reasons for the late start and rather slow growth of SMS usage. Firstly, the major cellular standards in the US (CDMA and TDMA) did not have SMS functionality. Secondly, paging has long been a popular service. Cingular, one of the biggest cellular operators, offers a two-way paging service called Blackberry (including a wireless email application) provided over a Mobitex network. The service was introduced in 1998 and by early 2002 it had attracted more than 800 000 subscribers (Lindmark 2002).

Compared to most European countries, the US enterprise market is still considered more profitable and interesting than the consumer market. Email messaging is more popular than SMS, postpaid payment solutions dominate heavily over prepaid, but wireless access via PDAs and laptops is more common. Email is currently the most interesting data service but it seems users find it difficult accepting wireless internet access with poorer quality than they are used to on the fixed net. In many cases users are experiencing very high roaming charges, especially if they subscribe to a regional operator. Also, many users are tied to the their current carrier due to payment schemes making switching to another operator difficult and often expensive. Due to lack of SIM card solutions on many terminals, many users experience a similar lock-in to their current device. Another feature hampering development is that many users have to pay not only for outgoing traffic but also for incoming traffic. This is due to the structure of the telephone numbers. There is no prefix telling a caller if they are making a call to a mobile phone or a fixed phone. To avoid unexpected fees for the caller, the mobile receiver is charged the extra cost. This problem is being investigated but it takes time to implement a solution.

Many new wireless services are aggressively marketed and implemented. Entertainment companies explore the wireless sector and are looking to partner with technology companies. Software for developing new applications and services (e.g., J2ME) are made available for downloading. The next two to three years will be critical for wireless services in the US. IDC predicts that by 2006 revenues from mobile data services will increase to $6.85 billion. Many of the problems discussed below need to be solved in order for this market to really take off.

Fragmented Operator Industry Being Consolidated

The US telecommunications industry was deregulated relatively early, in 1996. Today total wireless operator revenues can be estimated at about $65 billion, the absolute majority being revenues from voice services. The industry is fragmented with more than 300 wireless operators, most of them very small, operating on local and regional markets. There are six nationwide mobile carriers: Verizon, Cingular, AT&T Wireless, Sprint, T-Mobile, and Nextel. None offers anything close to nationwide coverage. The top three carriers, Verizon, Cingular, and AT&T Wireless, have a combined market share of 52% (Swedish Trade Council 2002). The carrier scene is becoming increasingly competitive and this has led to consolidation. Most carriers spend heavily on advertising. It is rumored that Verizon alone spends more than Coca-Cola on consumer advertising.

Multiple Cellular Network Standards

One important reason for the slower development of the US wireless market during the 1990s and early 2000s is the lack of a uniform air interface standard. There are four major competing 2G standards: CDMA, TDMA, iDEN, and GSM. Of the major carriers, Verizon and Sprint operate CDMA systems, the standard with the largest user base in the US. Cingular operates both TDMA and GSM systems. AT&T Wireless is changing its networks from TDMA to GSM and Nextel operates an iDEN system. Probably we will see two competing 3G standards in the future, CDMA2000 and UMTS/WCDMA. Six operators are currently rolling out different 2.5G infrastructures. At the same time, several old analog networks are still in operation with a rather large user base.

All these standards have resulted in relatively poor quality of service for the users, for example when it comes to roaming, coverage, and the possibility of easily switching between operators. With so many air interface standards and limited spectrum availability, it is difficult for the operators to deploy their network efficiently. Since GSM became the dominating global standard, instead of those originally chosen by the American operators, these operators could not enjoy the scale effects associated with large production volumes. In effect, the US cellular networks and the terminals became unnecessarily expensive. An example of the difficulties this situation poses for carriers and users is the decision recently taken by AT&T Wireless to change its infrastructure from TDMA to GSM/GPRS. Studies indicate that if AT&T Wireless had chosen GSM instead of TDMA from the beginning, its 2G network would have been about 25% cheaper to build.

WLAN: A Market Growing Rapidly

In contrast to these problems with the cellular infrastructure, we currently see a very rapid development, with the US as leader, in the deployment of WLAN networks in so-called hotspots. Currently there are about 15 million early adopters of the WiFi (IEE802.11b) standard in the US. The market potential is large, with about 45 million business travelers having a need for connections from their wireless devices while on the move (Swedish Trade Council 2002). By 2005 it is estimated that almost 14 million people will be working in the field, off site from office locations, creating an $8.4 billion market for wireless internet business applications and a $1.7 billion market for handheld devices (Swedish Trade Council 2002). The US is expected to remain the largest market for WLAN networks despite rapid deployment in other countries.

But problems remain to be solved here too. A particularly important issue is guaranteeing security in networks and transmissions. Other issues concern billing and the difficulties of providing roaming between networks, especially networks operated by different carriers and service providers. In addition, all devices are currently not equipped with WLAN capabilities and there are multiple technologies being deployed for WLAN (even if WiFi seems to win the battle) and little integration between them. At the same time, we see voice and data carriers cooperating or merging to complement each other's service offerings. Wireless internet service providers (WISPs) build hotspots and voice carriers supply billing solutions (e.g., the WISP MobileStar and the carrier T-Mobile).

Rather Weak Telco Vendor Industry

Compared to the dominating role of American IT vendors and developers, the traditional telecommunications vendors are rather weak from a global perspective. There are big American competitors in the cellular infrastructure and terminal markets, for example Motorola and Lucent, but they are not global market leaders. European companies, e.g., Ericsson, Nokia, and Siemens, dominate the infrastructure business. In the handset industry, the Finnish company Nokia is the market leader. Motorola comes second but it seems that the major contenders to Nokia's future dominance will come from Asia, for example Samsung, LG, and TCL.

The current integration of the telecommunications, data communications, and media industries and the increased deployment of WLAN networks in hotspots providing internet access and other data services, might lead to a stronger position for US companies. It seems possible that the winners will come from the IT industry, for example software companies, manufacturers of computers, routers, and other internet equipment, wireless access points, PDAs, etc., rather than from the traditional telecommunications industry. This might strengthen the position of the American IT sector even further.

Poor Coverage

The cellular systems in the US do not provide the same quality of service as in Europe and Japan. An important QoS parameter in Europe is the very good coverage due to the use of a common access standard (GSM) and extensive roaming agreements between operators. In the US, outdoor and indoor coverage is comparatively poor and fragmented due to lack of a standardized access technology but also due to various market inefficiencies and zoning issues, which make it difficult for operators to build new towers and base stations where they are needed. Not even the biggest wireless operators provide decent national coverage, and roaming is generally expensive for the user.

Lack of Spectrum Leading to Limited Capacity

Rapid increase in users of wireless services and limited spectrum availability have led to lack of capacity in many urban areas, resulting in poor service quality such as dropped calls. The spectrum issue is much debated and auctions of new spectrum licenses for 3G networks have been postponed several times by the regulating authorities, the Federal Communications Commission (FCC) and the National Telecommunications and Information Administration (NTIA). Therefore 3G licenses have not yet been awarded. Many of the frequencies that operators would need to increase the quality of service and offer new bandwidth-demanding services are currently controlled by educational institutions, the military, and television broadcasters. Carriers lobby for additional spectrum to provide more capacity and increase service levels. The FCC can be said to have a command-and-control approach to spectrum management. The previous auctions for 2G licenses were very complex affairs and the usage restrictions imposed by the FCC have created an inflexible system and a shortage situation. Only four carriers are licensed spectrum in a single market, increasing competition for spectrum, which leads to poorer coverage and high spectrum auction rates. Moreover, it was only in January 2003 that the FCC lifted a regulatory cap of 55 MHz on the amount of spectrum wireless carriers are permitted to possess in each urban market.

Another difference between the policy of the FCC and European regulators is that the FCC normally does not tie a specific technology to spectrum use. If an operator acquires a certain chunk of spectrum, that chunk can be used to deliver any type of service with any technology. In

Europe the 2G licensees have to use GSM and the 3G licensees have to use UMTS/WCDMA. The advantage of the European approach is that it leads to the emergence of a common and uniform standard with all the benefits earlier. The US policy, on the other hand, has led to several air interface standards but it also makes it easier for incumbent operators to smoothly migrate from one generation of network technology to the next.

In essence this has led to a situation where the American operators only have a few ways to deal with the capacity problems caused by the spectrum shortage in the short run. They can migrate to 2.5G or 3G technologies that use spectrum more efficiently. This is what we see happening today. Alternatively they can continue building 2G base stations to simply increase capacity in the existing networks. They may also engage in spectrum sharing with other operators or they can merge their operations to increase capacity, efficiency, and coverage. There are, however, signs that the FCC and other authorities are considering a more open and flexible spectrum policy to alleviate these problems. The FCC Spectrum Policy Task Force has been given the job of evaluating spectrum issues and the potential for deregulation. In February 2003 it was announced that the industry reached an agreement with the Pentagon to unlock a swath of spectrum for the next generation of wireless services. The new agreement will give companies access to 255 MHz of unlicensed spectrum around the 5 GHz band used in one of the WiFi standards for WLAN hotspots.

The Threat of Terrorism and Crime

A perhaps more intangible challenge facing the American communications industry is how the defense and other national authorities will handle the increased threat of international terrorism and other serious crimes. The possibilities of causing damage to people, society, and industry by using fixed or wireless communication networks are difficult to counter. It is not completely impossible that restrictions will be imposed on people's right to freely communicate over wireless and other networks. Another issue is the possibilities for the big content owners, many of which are large American media companies, to protect and gain revenues from their content being distributed over wireless networks. If these companies are unable to charge for services based on their content, development of the whole wireless industry will be hampered. Therefore problems relating to integrity, privacy, and security might pose substantial challenges, especially for the US wireless industry.

Europe

Western Europe is a larger economy than the US. Following enlargement of the European Union (EU) to include Eastern European countries, it will be the largest economy in the world. Europe is two years ahead of the US in mobile penetration, and GSM has given the European users a common standard with coverage, service quality, and global roaming. The European equipment vendors are world leaders.

The 15 member countries of the EU have a population of 370 million people. Ten countries will join in 2004 and two are pencilled in for membership in 2007. Including Turkey, which might start negotiations in 2004, this enlargement will add another 174 million people to the EU by about 2010. In addition there are 17 other sovereign states in Europe, among them Russia with 145 million citizens. All in all, the number of Europeans is close to 1 billion and the prediction is that in 2015 this figure will be about the same.

In 2001 the average GDP per capita in the EU was about $21 000, lower than both the US with almost $38 000 and Japan with $41 000 (Morgan Stanley 2002). However, the values for individual countries range from about $10 000 to $29 000. Initiatives such as the Euro and free movement of people and goods are in the process of creating a common European economy that is larger than the US and Japanese economies combined.

The GSM World Leader

The technological maturity of the Western European countries is high. Apart from the EU, what really unites the continent is the GSM standard. Whether a user is in Palermo, Hamburg or Helsinki, the GSM mobile phone keeps him or her constantly connected and reachable and always with the same number. Many Europeans have GSM mobile phones. Hungary, in the former Eastern, Europe, has a mobile penetration of 52%; Sweden and Finland, traditionally early adopters, have penetrations of almost 90% (Morgan Stanley 2002). GSM devices are used for more than talking. Text messaging, known as SMS, is a datacom service built into the GSM handsets for transferring text messages of up to 160 characters to another GSM handset, and this service is proliferating. Besides voice and text messaging, GSM can be used for mobile internet over the Wireless Access Protocol (Wap), but this has been a failure. Although GPRS (2.5G) can give Wap users higher bandwidth (32 kbps) and allows them to be constantly connected, Wap usage is very low. For instance, an early adopter such as Finland, with a population of 5 million and 90% mobile penetration, only has about 150 000 Wap users.

Problems with Seamless Mobile Access

For voice, GSM offers a seamless service across countries and operators. SMS is another stable service across the GSM networks. But for mobile data, the European GSM operators have failed on all levels. Users trying to use mobile data when roaming on other operator networks have seldom been able to get a connection. Before GPRS becomes a stable standard, each operator will be implementing it somewhat differently, adding complexity and confusion. Operators also failed to understand that mobile data should be preinstalled in new handsets and be activated simply by pressing the on button. Early users had to endure hours of complicated installation to get their Wap and GPRS connection working. Each handset model, and there are many, had its own flavor of implementing data access. The handset screens, navigation menus, and keyboards all looked a bit different from each other. This made it very hard for service providers to develop mobile applications that worked well for all handsets and networks. If the European operators fail in offering a seamless standardized platform and transparent billing schemes, Europe will lose in the mobile data market to Japan and other countries that manage to offer a service platform like iMode. Pan-European operators such as Vodafone and 3G operators such as 3 might show the way by enforcing a seamless service with branded handsets, etc. European operators and vendors need to become more user driven and less technology driven.

Telecom Debt Crisis

Anticipating an explosion in wireless internet business, European telecom operators staked unprecedented amounts on securing their right to a piece of the 3G spectrum. But converting those rights into profits will be difficult. The telecommunication giants of Europe have watched their stock prices crash at the same time as they owe well over $100 billon for the third-generation wireless license fees they paid governments in 2000. In Europe, for the telecom investments in 3G to become profitable, each customer has to spend about $50 per month. The current spending rate is approximately $25 per month. The operators hope their revenues will increase as the third generation of mobile telephony will lead to increased talk time, create more data traffic, and allow new services. But this is a tricky equation to solve.

Strong in Telecom, Weak in Datacom

Europe has world-leading telecom vendors but is weak in IT and datacom. Traditional telecom vendors have usually failed in their attempts to enter the IT and datacom market, which is dominated by American companies. The future prosperity of the telecom vendors is highly dependent on the survival of the telecom operators. If datacom technology takes over the telecom world, the European telecom vendors will face severe challenges.

Health and the Environment Taken Seriously

Health and environmental concerns are taken very seriously in Europe. The electronics industry is already obliged to take back and recycle sold goods. The high power consumption of electronic equipment will probably be addressed by the European authorities years before other parts of the world take any action. The issue of mobile phone radiation is already on the political agenda in Europe. In the case of a scientific gray zone (inconclusive evidence about dangers from cellphone radiation), European governments will probably follow the precautionary principle and regulate the wireless industry.

Stagnation and Overregulated Economies

Europe is a graying continent. A shrinking population segment of working age has to provide for a growing population segment of senior citizens. And nothing indicates rising birth rates. A further obstacle is the barring of economic migrants from entering the EU. Generous pension, unemployment, and welfare schemes are keeping healthy people out of the workforce, and this reduces economic growth. Strong labor unions add to the inertia in many countries. Deregulation of many former state monopolies is very slow. Increased time off for sickness, early retirement on pension, extensive vacation rights, and limited overtime working are reducing Europe's total working hours. Rigid labor laws and generous unemployment compensation make the mobility of the workforce minimal. Europe has problems competing with the US in science. Only if research politics are coordinated over borders will Europe be able to match the US. But that doesn't mean all countries can expect to be home to universities and research facilities of international renown. If nothing is done, then 50 years from now Europe could be an economic dwarf. The US and Asia will take the leading roles. A recent report from the European Commission shows that Europe in the middle of this century could be at a meager 12% of world trade instead of the current 20%. If this happens, the wireless industry will be seriously affected.

China

With a population of almost 1.3 billion and an economic growth rate of 8%, China is one of the most important economies in the world, taking a long-term perspective. In 2003 China is already the single largest market for mobile telephony. It has 200 million subscribers. The market for handsets is 70 million and there is a large domestic handset production. China's huge population means its long-term growth potential is immense. The Chinese government is aggressively supporting domestic infrastructure vendors such as Huawei. The company is already competing globally with Western infrastructure vendors in price-sensitive low-end market segments.

China has been aggressively using its position as the most important telecom market in the world when dealing with the incumbent telco equipment vendors, pushing for low prices and demanding access to technology. Competing Western infrastructure vendors are forced to overbid on promises of technology transfer. To gain entrance to the Chinese market, global handset makers have also been forced to form joint ventures with domestic handset manufacturers.

Due to the enormous size of the Chinese market, the winning standard in China will have a significant advantage in the global market. The Chinese government is using this for an aggressive industrial policy by promoting its own 3G standard, called TD-SCDMA, codeveloped by domestic company Datang. If TD-SCDMA becomes operational and China shuts out the global 3G vendors (WCDMA and CDMA2000), they will lose a large market and face competition from Datang in other international markets.

China has impressive competitive advantages in a huge domestic market. Among its assets are a long entrepreneurial tradition among Chinese expatriates, low labor costs, and a large talent pool of potential employees with academic degrees from top universities around the world. Companies in Taiwan and Hong Kong are integral as facilitators for the industrialization in mainland China. The political conflict between China and Taiwan has not hindered Taiwanese firms from making pragmatic business contacts on the mainland.

An Opaque and Overregulated Economy

China faces severe challenges in transforming and deregulating to unleash its full potential. China's economy is overregulated with an inefficient legal system. Civil law and company law need to establish property rights more firmly under an independent judicial system. An ever-present business risk during the current regime is the risk of arbitrary political interference from hard-liners in the ministries. Loss-making state-owned enterprises sub-sidized by the taxpayers are allowed to continue their operations, wasting valuable resources by producing inferior products. Lack of macroeconomic transparency adds the risk of hidden financial bombs that could trigger financial meltdowns similar to the 1997 Asian crisis. These factors are hampering economic growth. The establishment of reformed institutions and macroeconomic transparency are critical for the speed of future industrial development.

Political Instability

In China there is an additional challenge due to the risk of political up-heaval. The current ruling regime in China appears stable and in control, but better education and free information will eventually set forces in motion towards transparency, accountability, and the rule of law. Fundamental values in a population shift slowly. As a new generation evolves with better education, access to information and technology, and a higher level of material wealth, it will eventually demand political rights. This upcoming political transition could be gradual and peaceful, or it could be violent and throw China into a whirlwind of chaos. Due to China's size, reform movements will probably demand increased regional autonomy and a more federal state.

Risks of Complacency

The Chinese economy is growing by 8% per year. This impressive growth creates a risk of complacency over necessary reform. The risk is that the Chinese establishment starts to believe in some predestined Chinese in-vulnerability. Econometric studies indicate that the very high historic growth rates in the early NICs (Singapore, South Korea, Taiwan, and Hong Kong) could be explained by three traditional factors: transfer of labor from agriculture to manufacturing, growing labor participation rates, and high investments. Total factor productivity in these countries was comparable to OECD averages, not radically higher, as might be expected by the proponents of the "Asian miracle" hypothesis. Growth could then be explained by established macroeconomic theory without resorting to the "Asian miracle" hypothesis. As increasing total factor productivity is the ultimate driver of long-term economic growth, all NICs will eventually experience slowing growth. This will occur when the level of economic development approaches that of the OECD countries, and the transition to an advanced economy is completed (Oxford Analytica 2002). Therefore high nominal growth rates might turn attention away from the need for deregulation and structural reforms among the leadership in China.

Challenges for the Chinese Wireless Industry

Some of the Chinese challengers in the wireless industry, such as Huawei and Datang, will certainly be serious players in the future global market. Huawei's strategy is a prominent example. The company is aggressively scavenging for technological knowledge in hi-tech clusters around the world, and its entry strategy is similar to that of Japanese car makers a few decades ago. Huawei penetrates the infrastructure market by initially offering simple low-end products that target price-sensitive market segments. If Huawei succeeds in increasing its volumes, eventually it can improve its capabilities and target more complex and profitable high-end market segments. However, Huawei has to meet formidable challenges. The infrastructure market is a type of club, and all the operators have long-established relations with the incumbent Western vendors. The complexity of the telecom systems and the need for backward compatibility with the installed base of legacy systems are giving the established vendors a significant advantage in dealing with the operators.

After the breakup of China Telecom in 1999, China now has a handful of domestic telecom operators with various market focuses. The major players in 2003 are China Telecom, China Unicom, China Mobile, China Satellite, China Netcom, China Railcom, and Ji-tong. These operators have support from different parts of the government ministries, and the opportunities for international operators to attack this market are not very bright. Competition will probably be between the national operators trying to enter each other's markets. The large potential in the home market will make it less probable that these Chinese operators will pursue aggressive international expansions.

Risks and Opportunities with Chinese 3G Standard Wars

At first sight, the Chinese strategy of promoting a national 3G standard appears compelling. If successful, it will make Datang a leading vendor in the domestic market and a contender in international markets. Despite this, the forces of global standardization and scale economics are very powerful, and working against them is usually punished by the market. In the worst case, Chinese operators will find themselves locked into a proprietary 3G standard with higher prices, few vendors to choose from, and slower product development. Companies and countries usually lose if they take part in the standard fragmentation war and try to grab the entire cake for themselves. Europe chose one standard (GSM) for mobile telephony whereas the US fragmented into several incompatible standards. Today GSM is dominating the world market. Another less well-known example is the standards battle for digital TV in the mid 1990s. The US, with its very large domestic market, made a unilateral decision to chose ATSC as its standard. Europe reached a consensus around the more open DVB standard. Today almost all countries outside the US have chosen DVB and the vendors who bet on ATSC have a much smaller potential market.

Japan and South Korea

Together with China, Japan and South Korea are currently the most important markets for fixed and wireless telecommunication and data communication in Southeast Asia. Both have developed rapidly and are global leaders in wireless data services. Mobile phone penetration is quite high, 61% in Japan and 64% in South Korea (ISA 2002, Morgan Stanley 2002). The number of mobile subscriptions has already surpassed the number of fixed-line subscriptions in both countries (ISA 2002, Swedish Trade Council 2001).

The Japanese population is 127 million and the South Korean population is 48 million. Japan's GDP per capita is $41 000 and South Korea's is about $11 500. Both are mature IT nations. PC penetration is high on a global comparison, about 36% for Japan and 22% for Korea, but rather low compared to the US and many Western European countries (Morgan Stanley 2002). Whereas Japan's economic development has been very slow, even negative on occasions, since the beginning of 1990 Korea's development has been more positive. South Korea's economy has performed very well since the Asian crisis and is among the countries with the highest growth rates during the past 20 years. One of the main drivers behind this positive development is the telecom sector. Operators as well as terminal manufacturers have broken records in both growth and results. The export of telecom equipment amounted to $8 billion during the first half of 2002, replacing automotive as the biggest export sector. Korea is among the world leaders in providing its population with fixed broadband access and now has a household penetration of about 60%.

Leading the Way into the Wireless Future

Japan and South Korea play an important role in leading the global wireless industry out of the current recession. Japanese NTT DoCoMo launched the first commercial 2.5G wireless data service, iMode, in 1999. It has been very successful and in spring 2002 it had 32 million subscribers and boasted more than 3000 official sites and over 50 000 unofficial sites. In 2002 DoCoMo launched iMode in Europe, but with less success so far. Wireless data services have been very successful in South Korea too. In Japan and Korea over two-thirds of mobile subscribers are data subscribers, and they use it. Most use is related to entertainment such as character images, ring tones, games, horoscopes, and dating services. Mobile commerce is also starting to take off.

There are several reasons for the impressive development of wireless data services over 2.5G networks. In Japan, with a rather low PC penetration compared to the US and Europe, especially in homes, accessing the internet and sending emails from the wireless terminal is a substitute for the fixed net. Also, many Japanese people spend a lot of time on commuter trains and buses and they use wireless services during their journeys. The operators, especially NTT DoCoMo, are very dominant in the market, controlling the whole value chain, including networks, terminal design, content, and content aggregation. It is therefore possible to dictate the conditions, making sure that all pieces of the puzzle are in place for successful introduction of new services. But there are other important reasons for the success. Operators in Japan and Korea have a strong user focus in their wireless data operations. The services are easy to use. They have simple and well-functioning business models with revenue sharing between participating companies. There is a lot of content for the users and marketing has focused on functionality and branding rather than on technological revolutions. The handsets are not marketed as internet terminals, but as a way to access new and fun services. Prices for consumers are reasonable. All this means that users find it easy and interesting to try out new services and they gradually move to more and more advanced services.

Oligopoly in the Operator Industry

The operator industries have greatly consolidated in Japan and Korea during recent years. In Japan there are now three very dominat wireless operators. The giant is NTT DoCoMo with 57% market share, KDDI/AU with 25%, and J-Phone with 16% (ISA 2002). All three offer wireless data services over 2.5G networks. There is also a fourth but much smaller operator offering data services, DDI Pocket. The situation in South Korea is rather similar. As a consequence of the 3G licensing process, the five original operators have merged into three. The dominant player is SK Telecom with a market share of almost 54%; KT Freetel, the mobile branch of the recently privatized state-owned operator, has 32% of the market; and LG Telecom has 14%. There are essentially no virtual operators in Korea.

In both countries we see a strong oligopoly. This can be problematic in the long run if offers to consumers and prices become too similar and market positions for the companies are locked. This might lead to decreased competition and higher prices. Also, the existence of one very dominating actor on the market might be problematic. As an anonymous expert stated when commenting on DoCoMo's strong position, "DoCoMo is notoriously hard to work with, but impossible not to work with."

Multiple Standards for 2.5G and 3G

For 2G, Japan went down a path of its own, developing a separate air interface standard, PDC. Currently 2.5G services are offered over the packet-switched version of this standard (PDC-P) and over CDMA networks. South Korea, on the other hand, is a strong CDMA country. Both 2G and 2.5G services are offered over CDMA networks. For 3G, Japan has decided on two standards, WCDMA and CDMA2000. Three licenses have been awarded, one for each of the three major 2.5G operators. South Korea, initially awarded SK Telecom and KT Freetel licenses to operate 3G networks using WCDMA and later awarded LG Telecom a third license using CDMA2000. The existence of several standards can cause problems when it comes to interoperability, roaming, and cost for networks and terminals.

A Slow Start for 3G

The 3G licenses in Japan were awarded without auction or even a proper beauty contest. The operators only had to satisfy certain criteria. In South Korea the process was more complicated and much debated. Two UMTS licenses were auctioned in 2000. SK Telecom and KT Freetel agreed to pay about $1 billion each, half upfront and the rest in installments over a ten-year period. LG got a fairly similar deal for its license in 2001.

NTT DoCoMo launched the world's first 3G network in spring 2001; it covered Tokyo and Yokohama, initially for corporate customers. J-Phone and KDDI/AU started operations about a year later. Even though the operators did not pay any licensing fees, they face large investments in infrastructure to reach the goal of 90% coverage by the end of 2003. DoCoMo alone will invest about $9 billion in 3G through March 2004 (ISA 2002). In Korea, operators have so far primarily focused on developing and offering data services over 2.5G, but they are currently installing their 3G networks.

Despite what many had hoped for, the 3G services in Japan got off to a rather slow start. DoCoMo has problems with FOMA (Freedom of Multimedia Access) since the uptake among users has not been as fast as expected. The reasons are reported to be poor coverage, especially outside the largest cities, short battery life, buggy terminals, complex pricing, and expensive services. An initial lack of terminals was also reported. Some experts believe that most of these problems will soon be solved and that all 3G operators will roll out new services and offer new handsets during 2003. Two conclusions from this experience are that content is extremely important and consumer pricing has to be reasonable. The introduction of 3G and the possibility of higher data rates mean an opportunity for new and more advanced services. But these services will be very complex and the big challenge is to translate the 2.5G success into a similar 3G success.

A Saturated Voice Market

With the relatively high mobile penetration rates in Japan and Korea, the increase in new users is slowing down. In Japan the average revenue per user (ARPU) is very high, about $70–75 per month but it has decreased by 20% since 1998 (ISA 2002). Even though data-related ARPU, in for example iMode, is increasing rapidly, voice-related ARPU decreases even faster. ARPU in Korea is much lower, only between $20–25 per month, and has been stable since 1999 (Swedish Trade Council 2001). The trend is the same. Users spend less on voice services and more on data services. It seems that an important challenge for operators in Japan and Korea is to further increase ARPU for data services in order to pay for the very large 3G investments, upgrades, possible phone subsidy, etc., in the future.

3G Terminals Expensive to Subsidize

As 3G terminals have to be more complex to sustain the new multimedia services such as MMS, video telephony, streaming video, etc., they also become increasingly costly to manufacture, thus retaining a higher consumer price. One way for telecom operators to contract users to their specific services is by subsidizing the price of the terminal. This has been the case in both Japan and Korea. With increasing manufacturing cost, the subsidies for the terminals will have to increase to make them attractive for the end user. A common figure here is around 50% of the terminal price that the telecom provider will pay in order to tie a user to its particular services. This might put substantial financial strain on the operators, since 3G services can only be used with 3G terminals.

4G Already

In parallel with the rollout of 3G, research has begun on even more powerful mobile communication networks, also known as 4G. In Japan, NTT DoCoMo and others are currently investigating the possibilities of transmitting wireless data at speeds up to 100 Mbps. Japanese officials stress that 4G should not be seen as an evolution from 3G, but rather as a brand new technology. This research effort was started in 2001 and has caused controversy in other parts of the world, where companies perhaps feel they should reap the benefits of 3G before too much funding, time, and media attention are spent on the next generation.

No Major Infrastructure Vendors from Japan and Korea

Even though Japan and South Korea have very strong consumer electronics industries, neither has been able to gain a strong position in the global telecom market. In the handset market, the future looks bright for the Korean manufacturers LG and Samsung. They seem to be among the technological leaders and many experts have dubbed them the main contenders to Nokia. While other manufacturers release 15–20 new terminals per year, Korean companies release half a dozen per month. The infrastructure market is even worse. The relatively weak position of Japan and South Korea might be surprising at first glance, but one important reason is that neither country chose GSM, the 2G standard that won the market war. Most players dominating the infrastructure market originate in GSM countries except for Alcatel and Lucent, neither of which is a market leader. With both Japan and Korea choosing WCDMA as one of their 3G standards they probably hope for a stronger market position in the future. But they will face tough competition from the established companies.

The Japanese Recession

South Korea has had a very positive economic development following the Asian crisis in the 1990s, but Japan is struggling with an economy that will not take off. Since 1991 Japan has gone through three recessions. Consumer prices have not increased since 1998 and currently the country is caught in a negative spiral of deflation, hampering development. If this spiral is not broken and if the required structural reform is not put in place, Japan's economy might not get back on track in a long time. This will seriously affect not only the development of the wireless industry and wireless technology but also the region as a whole.

Political Uncertainty on the Korean Peninsula

The conflict between North and South Korea is a constant threat to the region and its markets. Increased conflict will most definitely have a negative impact on the development of the wireless industries, especially in South Korea, though it is hard to predict how and by how much.

Part IV

Moving into the Future with Scenarios

11

Scenario Thinking and Scenario Making

This chapter describes a method and working procedures for creating the scenarios and formulating the key research areas. We briefly describe logics of scenario creation, referring to work by the highly influential Peter Schwartz. We describe how we created our scenarios and we conclude with an overview of some other scenarios and future-oriented studies that have been important as input and inspiration for our work.

Logics of Scenario Creation

Many logics of scenario creation that we used when writing this book were inspired by the work of Peter Schwartz, the author of *The Art of the Long View: Planning for the Future in an Uncertain World*. Apart from being a prominent writer, Peter Schwartz is cofounder and chairman of Global Business Network. An internationally renowned futurist and business strategist, he is a specialist in scenario planning. Schwartz works with corporations and institutions to create alternative perspectives of the future and develops robust strategies for a changing and uncertain world. We have found Schwartz's writing a prolific source of methods and ideas on how to conduct scenario work and strategic conversations. We have adapted his suggested techniques to our specific needs.

In scenario development we create and then reflect deeply on stories of probable futures. The stories are carefully investigated, full of relevant detail, leaning toward real-life conclusions, and hopefully designed to bring forward unexpected and sudden leaps of understanding. The scenarios can be seen as a tool for ordering our perceptions. The point is not to create your preferred future or to hope it will materialize, although there are some situations where acting to create a better future is a useful function of scenarios. Nor is the point to find the most probable future and then try to adapt to it. The point is to make strategic decisions that will be sound for all probable futures. No matter what future takes place, you are much more likely to be ready for it and influential in it if you have thought about scenarios thoroughly.

The scenario process provides a context for thinking clearly about the impossibly complex collection of factors that affect any decision. It gives us a common language for talking about these factors, starting with a series of "what if" stories. Furthermore it encourages participants to think about each of them as if they had already come to pass. What if our worst nightmare took place and our primary business became obsolete? Or what if our most desired future came to pass? What unexpected challenges would it present to us? Or what if a completely unexpected series of events changed the structure of our industry? Would we be overwhelmed or would we see the opportunities?

Thinking through these stories and talking in depth about their implications bring to the surface people's unspoken guesses about the future. Scenarios are a powerful tool for challenging our mental models about the world and lifting the blinds that limit our creativity and imagination.

Our Approach: Trends

The method we used in creating the scenarios is in many ways similar to what is usually recommended by other scenario makers and theorists. The most important difference is that in the early creation process we did not identify two main variables as a foundation for the final four scenarios. One advantage with an approach with two main variables is that you can easily map the created scenarios to an illustrative 2×2 matrix. A problem is that you must know and decide which two variables you think are the most important ones. All other variables then become more or less unimportant. Another is that they have to be independent of each other, and often they are not. Instead we developed the scenarios by combining 14 identified trends in different ways, giving us more freedom in their creation. On the other hand, it is a bit unwieldy to illustrate a 14 × 14 matrix on paper.

Driving Forces: What Do We Care About?

Every enterprise, personal or commercial, is propelled by particular key factors. Some of them are within the organization, for example your workforce and goals. Others, such as government regulation, come from the outside. But many outside factors are not completely obvious. Identifying and reviewing these fundamental factors is the starting point and one of the objectives of the scenario method. In other words, driving forces are the elements that move the plot of a scenario; they determine the story's outcome. Without driving forces, there is no way to begin thinking through a scenario. They are devices for sharpening your initial judgment, for helping you decide which factors will be significant and which factors will not.

Fundamental Drivers: What Do We Know?

Fundamental drivers are trends with a reasonably high probability of coming true in all scenarios. They are not contingent on a particular chain of events. If it seems certain, no matter which scenario comes to pass, then it is a fundamental driver. Identifying fundamental drivers is a good confidence builder. You can commit to some policies and feel reasonably sure about them. There are several useful strategies for investigating fundamental drivers:

- *Slow-changing phenomena*: these include the building of infrastructure and the development of resources. The most commonly recognized fundamental driver is demographics. For instance, as soon as the postwar industrialized countries' baby boom began, it was obvious that the babies would eventually age. The effects of this aging are still uncertain.
- *In the pipeline*: today, for example, we know almost exactly how many teenagers there will be in Europe in the 2010s. All of them have been born therefore all are in the pipeline. The only uncertain factor here is the amount of immigration.
- *Technological growth paths*: we know that the performance of computers has been growing at the rate of 55% per year since 1940, popularly stated in the 1960s as Moore's law. After talking to experts in the field, we can assume with reasonable certainty that this growth will continue over our planning horizon.

Critical Uncertainties: What Do We Not Know?

In every plan there are critical uncertainties. Although we might know the approximate numbers of a potential market, we don't know how many in that market will be sufficiently well off to buy a given product. Critical uncertainties are intimately related to certain fundamental drivers. You find them by questioning your assumptions about fundamental drivers: What might cause the price of wireless access to rise? What might a large wireless provider do to retain its domination over the voice telephony market? If traffic jams becomes too heavy, will there be a change of habit? Would businesses introduce very flexible working hours? Would they allow tele-commuting? Would real-estate prices drop enough so that people could afford to move closer to work? Or would onboard car computers, automatic navigation systems, and wireless telephones turn the automobile into a mobile office? We cannot know for sure and any look into the future of traffic jams could include at least two scenarios: one where many cities are paralyzed and one where commuting undergoes a transformation.

Making Our Scenarios

The standard method of scenario development used by large consultancies is a very structured process. It is derived from their prevalent hypothesis-driven, high-speed working method. This method is built on quickly identifying what's most relevant, cutting the ambiguities, and acting quickly on the relevant information.

The mainstream structured scenario method usually starts with an open brainstorming phase for generating as many trends as possible plus other factors that affect the topic under discussion. After that, all identified variables are rated by importance on a number of different dimensions. From this, factors of lesser importance are dropped from the analysis, reducing the dimensionality of the scenario space. Variables that are dependent and closely correlated are checked to see if they can be merged. In this way, the complexity is reduced in a number of iterations when less important scenario dimensions are dropped. Ideally, you end up by identifying the two most important dimensions for the subject. As these two dimensions are supposed to be independent, you can illustrate the scenario space using a 2×2 matrix. Finally, four scenarios are formulated, each in one corner of the matrix.

This method is well structured and the 2×2 summary with four scenarios is very easy to understand. However, this hard-structured method has a number of drawbacks. The most important is that you are forced to formulate all scenarios along the two final dimensions. This is fine if there are two dimensions that are much more important than all the other identified variables. These dimensions should ideally be independent. If not, the method itself will force you to ignore dimensions that could potentially be important in the future.

During our process, we were striving to keep the method and format for the final scenarios rather open. When formulating the four final scenarios, we chose to explore what we believe are important topics for the future of the industry. We did not have the goal of reducing the four scenarios into something even more fundamental, like a 2×2 matrix. In an n-dimensional space, our scenarios point in different directions.

Creating the Scenarios and Key Research Issues

The work on developing the scenarios was conducted in an iterative manner. Between weekly work meetings the team members worked independently on agreed tasks. An important issue was creating a common framework and understanding among the authors, who came from diverse disciplines. The common framework was created through a series of meetings where current topics were discussed, including the working process and scenario methodology. Another important aspect was the sharing of knowledge about the telecommunication industry, its markets, technology, user behavior, etc. The meetings were complemented by studying many existing reports, books, articles, relevant future predictions, etc.

The next step was to identify fundamental drivers of development, or mega-trends; it involved taking a first step in understanding what is driving the wireless industry, technologies, users, etc. Initially these drivers were divided into four broad categories: (1) technology, (2) socioeconomic and political, (3) business and industry, and (4) users, values, attitudes. The list was continually iterated in a long process of discussions. The fundamental drivers were subsequently assumed valid for all scenarios, forming the basis for each of them.

During this phase we organized Wireless Round Table I with external experts. The experts invited to this seminar came from academia as well as from industry, covering a wide area of expertise. From academia, experts on future studies, lifestyle studies, and telecommunications participated. From industry the experts represented operators, service providers and developers, infrastructure vendors, venture capitalists, bankers, and consultants. The industrial participants represented their personal areas of expertise rather than their companies. The intention with the wireless round table was to openly discuss the possible evolution of the wireless world and to identify key areas of agreement and disagreement. The round table was led by a moderator and started with four of the experts presenting their personal views of the wireless world in 2015, serving as an input to the following discussion. Ending with a survey distributed among the experts, giving feedback on their views of the wireless future and the driving forces affecting it, the meeting was an excellent source for expert input.

The work continued with identification of a large number of trends in the wireless world from the fundamental drivers, forming the basis for developing a first set of scenario embryos. During a long and iterative process, 14 trends were selected as especially important. Moreover, these trends were considered uncertain in the sense that their direction and impact were vague. The trends were then used as dimensions differentiating the scenarios. Initially about a dozen scenario embryos were formulated. Gradually the embryos were merged or discarded, culminating in four scenarios that were further refined over time. During this phase we organized Wireless Round Table II with external experts. This time 11 experts from various fields participated. The objective was to get concrete feedback on the four scenario drafts. Two similar workshops were later organized with half a dozen representatives from the academic world, i.e., professors, researchers, and engineers doing research and development on wireless technology or related fields.

In parallel with fleshing out the scenarios, we began formulating the so-called technical implications from the scenarios and the research challenges. This phase involved taking the four broad scenarios and seeing what they imply for technological research, formulating assumptions about the wireless future. Starting from the technical implications, highlighting technological and other bottlenecks that might hamper a positive development of the wireless world, we formulated a set of key research areas.

Finally, we began identifying challenges for the wireless industry and for the most important nations and regions over the next decade. Here the main input was literature and articles, a few studies commissioned and performed by external specialists or students, and interviews with experts.

Weak Signals and Provocative Questions

The process described above perhaps looks quite rational and straightforward. But in practice the work was conducted in a very iterative manner, with the phases carried out in parallel. On numerous occasions we went back and made alterations and changes to work in progress that had been left to mature.

Nevertheless, this approach is traditional in the sense that it begins with the world as it looks today and then identifies driving forces and trends in an attempt to say something about the future. It is quite similar to the traditional way of predicting the future based on trend extrapolation. In both cases the starting point is the present.

As a complementary approach, we tried to start from the other end as well, trying to put ourselves in 2015 looking back. We did this by posing provocative questions and looking for weak signals. This approach was fruitful in freeing our thinking from the boundaries of the present. Here are some examples of provocative questions:

- How would the wireless world look if base stations could be bought and installed by any user at a very low cost and the user could earn money from providing wireless access to others?
- What if radiation from mobile terminals proved to be harmful to humans after lengthy exposure?
- What would happen if one or more of the world's large service providers for wireless access went out of business due to large debts?
- What if the media industry lost such large amounts of money due to piracy that they were forced out of business?
- What if hacking became such a big problem that users no longer trusted sensitive wireless services like banking or payments?
- What if security issues meant that governments decided not to release more spectrum?

Information and Feedback

During the process, work in progress was presented and discussed on several occasions in different environments. The two most important external events were the two round-table seminars and the two academic workshops. We were also invited to present the scenarios and trends at several workshops and conferences, and they too provided good feedback. A large number of experts from different fields were interviewed, giving their views on the wireless future.

Commissioned Studies

Outside specialists performed specific but small studies. To learn more about the most important NICs, two studies on the wireless markets, industries, and actors in India and China were initiated. These studies, performed independently by one expert from India and one from China, were very valuable for understanding the development potential in these large nations (Bhattacharya 2002; Long 2002). The authors also initiated two M.Sc. projects—one on the mobile services market in Russia (Sahle and Thuresson 2002) and one on mobile multimedia services over 2.5G (Schildknecht 2002)—and three smaller student projects. Two of the smaller projects focused on different aspects of WLANs and WISPs (Arvedson, Edlund, and Thomsson 2002; Bergqwist, Engren, and Eriksson 2002) and the third looked at future spectrum requirements (Etemad and Lennerfors 2002). These studies provided us with valuable and detailed knowledge about specific aspects of the wireless future.

Other Studies about the Future

Even a very limited literature search shows there are many scenarios out there. Adding all unpublished scenarios made by companies in their strategic work increases that number even more. Here we briefly describe a few studies about the future of the data and telecommunication industry. Our aim is to present the studies that have influenced us most in our work. We outline the most important features of each study and highlight some of the similarities and differences with our scenarios.

The PCC Research Program

In the project Fourth Generation Wireless (4GW) of the Personal Computing and Communications (PCC) research program, scenarios have been used as a tool for formulating research topics related to future wireless systems. By working with scenarios, the 4GW project group tried to challenge some of the implicit assumptions of their research traditions. The vision of the 4GW project is "Personal Multimedia to everyone at today's prices for fixed telephony" (Bria et al. 2001).

The approach used in the 4GW scenario started with creating technosocioeconomic scenarios based on literature studies, interviews with experts from academia and industry as well as a so-called Delphi study.

From the basic assumptions, a number of working assumptions were formulated. They represent the more operational goals of the research program and have been used to formulate the specific research topics of the 4GW project. The working assumptions about 2010 were (1) telepresence will be an important driver, (2) information will be available anytime and anywhere, (3) intermachine communication will be an important service, (4) security will be an indispensable feature, (5) one-stop shopping services will be common, (6) the infrastructure will be nonhomogeneous, (7) public and private access will be common, (8) ad hoc, unlicensed operations will dominate, (9) multimode access ports will be common in the public systems, and (10) terminals will exhibit a large range of bandwidths.

Three scenarios were formulated: Pocket Computing, Big Brother, and Anything Goes. They address issues like user behavior and lifestyle, the evolution of the telecommunications market, development of supporting technologies, and the evolution of values and society.

Our work to some extent builds on the project carried out by PCC. Several of its working assumptions are similar to our technical implications drawn from the scenarios, but labeled differently. A difference is that our view of the future of the wireless industry is broader and placed in a richer environment.

The WWRF Book of Visions

The Wireless World Research Forum (WWRF) is an organization for industry and academia active within the wireless field. It publishes *Book of Visions* (WWRF 2001), a collection of ideas, visions, and important research topics for the future of the wireless industry. Four working groups focus on (1) the human perspective; (2) service architectures of the wireless world; (3) new communication environments and heterogeneous networks; and (4) spectrum, new air interfaces, and ad hoc networking. Each group proposes several research tasks within a time frame of 10–15 years from now. Participants mainly come from industry but lately the involvement from academia has increased substantially. The declared approach is to not formulate purely technical visions, but to use a broad context when describing the future and to take a user-centered view when considering new services and business models. With this work WWRF also tries to create a shared vision among the important players in the wireless field.

The large and expert production team for *Book of Visions* allows the proposed research tasks to be formulated in more detail than in our study, even detailing specific research packages. They look directly at the technological development, put it in relation to the user, and then formulate their visions. These visions are then developed into complete and detailed work packages. We have put visions, trends, and fundamental driving forces together to create broad scenarios and then identified technical implications and research issues. Since many of the research issues proposed by WWRF are found in our report as well, we can say that our scenarios support their visions.

Swedish Technology Foresight

Swedish Technology Foresight is a very large study carried out by the Royal Swedish Academy of Engineering Sciences (IVA). The effort covers a large spectrum of society, industry, and technology. Examples of sectors are healthcare, biological resources, society's infrastructure, and education. The area most relevant to us is information and communication systems (IVA 2000).

The purpose of the information and communication systems study was to create knowledge and visions about the technological development within the coming 5–10 years. By formulating visions about future society, seven key areas of technology were identified: (1) always connected, (2) the digital assistant, (3) a growing use of software, (4) future services are electronic, (5) ubiquitous and instant learning, (6) the technological and biological worlds meet, and (7) security and integrity. These areas will have a significant impact on information and communication systems during the coming decade.

The seven areas were formulated by analyzing fundamental drivers and inhibitors, both in Sweden and globally. Five important driving forces were identified: (1) increased globalization and dissolving of national borders due to economic, cultural, and technological trends; (2) convergence of technologies, functions, products, services, and markets; (3) a more positive attitude towards entrepreneurship and venture capital, in particular in Sweden; (4) individualization of society, leading to new products and markets; and (5) issues around regulation and deregulation of the information and communication technology markets. The report ends with a discussion of the strategic consequences for Sweden as a nation.

Our work is both similar and different. Both are broad in scope, taking societal development as a starting point, but no actual scenarios are developed in Swedish Technology Foresight. Most of the drivers it identified are included among our fundamental drivers. Swedish Technology Foresight focuses on development in Sweden whereas our work takes a global perspective. The most important point is that we have used the seven key areas of technology as an input to our work. They are well in line with some of the key research areas we recommend, even though we use a different categorization and other labels.

Beyond Mobile

The study *Beyond Mobile* looks at the future mobile marketplace with a time horizon of 2005–2007 (Lindgren, Jedbratt, and Svensson 2002). It does not put much emphasis on the technical aspects of wireless development, but concentrates on commercial and business aspects. It is written from two main perspectives, human and business. These two perspectives are then placed into the two contexts that the study uses to discuss development.

The study is divided into four main sections: (1) context discusses the fundamental dimensions of what a mobile society means for the consumers and the marketplace; (2) drivers and trends discusses changes in technology, institutions, humans, society, and the economy then maps them onto the development of a wireless future; (3) arena pictures the future mobile marketplace mainly from an industry and business perspective; and (4) the mobile marketplace of tomorrow discusses possible consequences and futures. There are also four very short scenarios of possible futures.

With its market perspective on the future mobile marketplace, *Beyond Mobile* focuses on society, humans, and economic theory rather than the technology behind it. Though we haven't put much emphasis on the four scenarios besides reading and discussing them, the study's thorough discussion of softer developments in society has been useful and inspiring. Perhaps the most useful part of the study is the definition of the future user segments; they are very well defined and have matched our own research in this area.

Other Scenarios

During our work we have had access to several proprietary scenarios from large industrial players. These scenarios have been developed as part of their strategy process to map long-term industry development. They have been helpful as a point of reference on structure and method, and to some extent for their content. Scenarios developed for internal use are generally more limited, mainly focusing on what's immediately relevant for the company. We have also read a large number of other scenarios and future-oriented studies. Here are a few studies that have been valuable but not explicitly used: 3 (2002), WWF (2002), IST (2001a, 2001b), and McClelland (2002).

12

Summary and Concluding Remarks

In this final chapter we summarize the main themes and discussions in the book. The first section starts with short summaries of the four scenarios and continues with descriptions of fundamental drivers, trends, and conclusions drawn from the scenarios. It ends with a summary of the most important challenges for the future facing the research community, industry, and key regions of the world. The chapter concludes with a brief reflection on the importance and the difficulties of taking a step back from the present when trying to envision the future.

The Book in Brief

Four scenarios describing how the wireless world might evolve from the present to 2015 constitute the core of this book. The scenarios are concrete images, including descriptions of the wireless systems of 2015, how these systems are used, and who are the most important actors and users. The main focus is the challenges and development of the wireless industry, i.e., operators, infrastructure vendors, terminal vendors, service providers and service developers. The scenarios are not intended as predictions but as a source of inspiration when thinking about the future of wireless technology and the wireless industry.

Wireless Explosion—Creative Destruction

Wireless applications and services are a huge success in 2015, and in a rapidly transforming industry the old market leaders have lost their dominant positions. The old telco world with closed, vertically integrated solutions gave way to layered, open architectures based on the Internet Protocol (IP). The datacom industry won the market battle. However, in a large but maturing industry, profit margins were squeezed and the datacom winners could never really leverage their market power. The wireless success changed people's working habits and lifestyles. Being always connected, a large part of the workforce can spend most of their time on the move, in meetings or traveling between meetings.

Rapidly growing industry

The economic downturn in the early years of the century slowed industry growth for a few years. However, the rapid technological development within the communication and information technology industries continued and essentially all markets and industry segments experienced more or less continuous growth.

Industry fragmentation—market leaders losing hegemony

The incumbent players consolidated but in a maturing industry profits were eroding as the products became low-margin commodities. Independent consumers undermined IPR enforcement. Open-source software and do-it-yourself wireless access further undermined corporate hegemony. The dominant market leaders did not vanish but the rapid technological development began turning profitable products into low-margin commodities. Industry fragmentation and vertical disintegration accelerated when companies became more and more specialized. When performance of any given technological function was good enough, design and manufacturing knowledge was no longer a critical asset and modularization set in. As a consequence, this part of the market split into several new markets.

Debt-burdened operators losing market dominance

When wireless data started, traditional operators first tried to offer closed telco-style services and developed in-house wireless portals. Seamless roaming, as in the voice GSM world, was very hard to accomplish with wireless data over a number of different underlying networks. The operators failed. The major blow to operator dominance was the rise of unlicensed spectrum and WLANs.

Telco equipment and terminal vendors lose to datacom attackers

Traditional telco equipment vendors failed in responding to the disruptive innovations. They were adapted to a business model built on selling extremely expensive systems to a few very demanding operators and they were dragged down together with their traditional customers. When the market fragmented, attackers captured emerging submarkets such as base stations. Telco terminal vendors lost market power when commoditization of the market occurred. The critical telco knowledge embedded in the radio and codec software was eventually commoditized by attackers from the NICs and the datacom industry.

An explosion in services and applications

In the industrial countries and the most successful NICs, cellular systems are complemented with a large number of other systems (e.g., ad hoc networks, WLAN access, satellites, high-altitude platforms). Most problems concerning seamless roaming, system integration, etc., have gradually been solved. The appetite for wireless applications and services is very high and once the new geographical positioning infrastructure was in place, the number of location-aware services grew rapidly. Wireless services are used by everyone.

Spectrum—abundant release for unlicensed bands

During 2005–2010 governments released significant chunks of new spectrum. With much more available spectrum, traffic prices fell rapidly and the dominance of the incumbent operators was reduced. Unlicensed spectrum usage was a huge success. The unlicensed bands drove rapid innovation of cheap install-it-yourself black-box access points that can double as multi-band base stations.

Batteries and complexity management no showstoppers

The lifetime of batteries for mobile terminals has increased dramatically since the turn of the century. Batteries are now used on a large scale for an enormous number of services, which has led to large production volumes and price drops for these new energy sources.

Slow Motion

The wireless world has developed slowly since the turn of the century. The global economic recession during the first decade in combination with real and perceived health problems due to radiation from wireless devices deeply affected the industry. Even though the demand for mobile services has increased, the service explosion that many people envisaged never materialized. The wireless industry has gone through substantial change. Consolidation has increased and the number of companies in each market has been reduced. Technological development has slowed down and profit margins have decreased substantially. The industry has matured. The big NICs, for example China, India, and Russia, are catching up faster than expected.

Economic recession and 3G fiasco

The global economic downturn that started in 2001 turned into a large-scale economic recession. The telecom, computer, and media industries were severely affected. It became really bad when a large European operator went bankrupt. This spread very quickly to other operators and eventually to vendors and service providers. Several large telecom actors disappeared and those that survived made massive cuts and saw drastically reduced margins. Many 3G commitments were renegotiated. Some networks were canceled and many were merged, resulting in only one or two networks per country. In many rural areas there is still no 3G coverage.

Health problems from radiation

The long-term studies of how radiation affects humans, presented around 2005, still have a negative impact on the industry. The results were clear and most experts agreed that wireless devices, when heavily used, would injure the brain due to radiation from the transmitter. In the beginning, the telco industry argued that the results were inconclusive, but eventually adopted a proactive strategy and managed to avoid total disaster by suggesting strict regulation of radiation levels and by redesigning their products. Usage is still affected, even though most problems are solved.

Security a problem still waiting to be solved

The problem of hacking and virus creation is still significant. Most security codes are quite easily broken and viruses are spread in the wireless networks. The problems increased when data services were introduced in the updated 2G systems and were further accentuated with the introduction of 3G. Many people feel that they cannot trust electronic transactions and are seldom willing to e-shop.

The mobile lifestyle loses ground

In the Western world and in Japan the mobile lifestyle came to a halt during the first decade of the twenty-first century. Many people, especially young families, moved from the cities to smaller communities and working from home or in local offices became increasingly popular. The result is that fewer people travel long distances to work. One important driver behind this shift is increasing environmental awareness. Environmental groups also started to campaign for decreased use of communication devices. For some time, usage was negatively affected but eventually industry was able to handle this issue by significantly reducing the power consumption in equipment and devices.

No service explosion

Despite the hype at the beginning of the century, the mobile service market has experienced slow growth. Most services used by consumers are still quite simple, focusing on satisfying basic communication and information needs. Many consumers are simply not prepared to pay for advanced services at the price they are offered.

Wireless telecommunication is a mature industry

Wireless has become a mature industry that has gone through consolidation and restructuring. The technological development has slowed down considerably and profit margins in all sectors have decreased substantially. Many platforms, solutions, and components are still designed according to closed and incompatible standards protected by patents. Concentration has increased and the number of players in each market is rather small.

The big NICs catching up after a slow start

The slow development in the Western world and in Japan in the first half of the 2000s was reinforced by problems in the big NICs (China, India, Russia, etc.). However, around 2010 the situation had improved substantially in many of these countries. Investment in infrastructure started to increase, giving the vendors a chance to recover some of their declining sales. The big NICs are now by far the most important markets for systems and terminal vendors. Moreover, there are now important global players such as operators, vendors, and service providers based in some of these countries.

Power consumption and complexity management as technical limitations

Despite large research efforts on new battery technology, no significant progress has been made. Many advanced services are almost impossible to run when the terminal is on battery power. Despite the slow development, several different types of system exist. Cellular systems of different generations coexist with other types of system (WLANs, PANs, broadcasting, etc.). But the problems of managing this complexity in a seamless manner are yet to be solved.

Rediscovering Harmony

Balance in life became the dominating value in most industrialized nations where material abundance could be taken for granted. These are post-materialistic times where human and environmental needs are in focus. The wireless industry is experiencing a difficult dilemma: refocus or die. There are fewer service and application providers than predicted around 2000, but the market is not completely dry. The big hurdle is to refocus and rethink business models, offerings, and brand in a market with active and demanding consumers categorized by numerous subcultures with individual needs. We see many local operators and service providers that have emerged as a result of the trend for people to move out of the crammed cities, forming smaller, local communities. At the same time, there are a few global operators providing global communication for the increasing number of people traveling longer and more often for pleasure, and for smaller but more price-insensitive segments.

A sustainable society in balance with itself

The industrialized world is based on the idea of a sustainable lifestyle where friends, family, and the environment are key elements. The high-paced lifestyle that dominated the Western world in the closing decades of the twentieth century finally went out of control. Consumers became more and more indifferent to brands and commercial messages and no longer accepted companies ignoring ethics, environment, human needs, and product quality. As a result, we saw a number of movements that combined a more sustainable and human perspective on society with a strong individual and social focus. To consider the environment and human needs has become valuable in the marketplace.

Two market segments driving the development

The move towards the new lifestyle started in two segments: Moklofs and Elders. The Moklofs are strongly focused on entertainment and messaging services. They participate in communities, both local and global, and are very global in their ways of thinking. They are open-minded but they don't believe smart marketers trying to claim that they will get a new life by buying the latest gizmo. Living in a world of tribes with many lifestyles, they want to express their affiliation with clothes, looks, and stuff they use. The Elders place high demands on usability and quality of service and they are not afraid of letting their voice be heard. It is very important to communicate with the family while on the move or when living apart. Healthcare is another important segment, allowing people to check up on their health wherever they are.

Less but more travel

People are moving out of the crammed cities and into smaller and cleaner communities in the suburbs or countryside. The lifestyle trend is to live and work in small, local, and very social communities. In the cities, the public transportation systems have been upgraded and the number of cars has decreased. This is due to harder environmental laws and political decisions to turn more of the city areas into car-free zones. Leisure travel is the only form of travel that is increasing all over the world, creating a demand for global communication possibilities.

A few clouds in the sky

Health risks and integrity problems are widely debated, but it is the telco industry's impact on the environment that people are most concerned with. Brominated flame retardants used in electronic equipment have turned out to be damaging to the environment and humans. Power consumption for terminals and infrastructure is another issue where consumers want to see improvements. Health threats, real or otherwise, are hard to battle, forcing the telco industry and governments to find new ways of restoring public trust in wireless technology.

The industry dilemma: refocus or die

After the initial wave of excitement over the new communication possibilities with 3G, the pace of development slowed down. This left the telco industry confused. The main reason was the industry's inability to adjust to the mass market's new attitudes and values. The industry is currently regrouping and adjusting to the new situation. Some players changed their business models to suit the new fragmented marketplace and became highly successful. Other companies failed to understand the new environment. The big hurdle is to refocus and rethink business models, offerings, and brand.

Peer-to-peer applications and services a hit

Despite the changes there is still a demand for wireless services but the main difference is that the mass market is selective about what kind of information is being received and when it is delivered. A new market has gradually emerged where personalized and very specific types of services are successful. Examples are peer-to-peer communication services, multimedia messaging, location-based services supporting social interaction, and devices and services forming family intranets. The demand for peer-to-peer technology has led to a fierce debate on how to solve the problems with IPRs. The content providers feel that the operators don't take responsibility for how their networks are being used, whereas the operators argue that they are simply providing the infrastructure for communication.

Big Moguls and Snoopy Governments

Through consolidation and mergers, large companies, known as moguls, have come to dominate the market. A mogul is a descendant of the early big information technology or media companies that managed to survive the crises of the first decade in the twenty-first century. These moguls grew and expanded outside their original business segments; for instance, from being only a systems software manufacturer one company became a big content provider and also started manufacturing devices aimed specifically at its own services. Smaller players were often bought or put out of business due to the dominant position of the big companies. The moguls, together with the world's governments, exert substantial and active control over the information flow and the communication industries. The purpose is to protect society and individuals from various unwanted actors and behavior, such as cybercrime, international terrorism, and illegal copying of software, music, and movies. Anonymity on the net is no longer possible. All users are automatically identified and registered when acting on the net. However, the world is not an antidemocratic society where the net is used to gain power and ultimately dictatorship, even though many people fear this might be the case. Counter and freedom movements do exist, despite heavy measures against them by governments and large corporations.

Moguls and governments

In each market segment there are now only one or two totally dominant market leaders. Some market leaders have been able to expand their market power into other areas. Users like these big companies because they feel they can trust them and their products fulfill their needs. There are no longer compatibility problems with software and hardware as there is only one choice. Governments like the big companies since they think they are easy to control. To some extent, the moguls agree to this control, as long as the governments do what the moguls want.

Security problems of the 2000s solved

Governments and industries took strong measures against the security problems of the early 2000s. In 2007 the first secured devices were introduced by one of the major hardware and software developers. These devices relied on new, "unbreakable" encryption technologies and required a personal certificate plus user biometrics. They contained circuitry for monitoring traffic and sending information on possibly unapproved traffic to government agencies.

Moguls in control and slow development in the NICs

Network effects, economies of scale, and successful enforcing of intellectual property rights created a new global economy with large players becoming even larger. The US government abandoned the antitrust laws of earlier centuries, allowing already big players from America to grow huge in the global market. Even though there were quite a few positive signs in the big NICs (e.g., China, India, and Russia) in the early years of the century, their difficulties continued. China's integration into the world economy slowed down due to political instability.

Incumbent telco players keep control of the market

With traditional mobile operators dominating over new actors, the strategic success factors proved to be brand ownership and customer relations. The leading European operators managed to survive the financial problems in the early 2000s through debt restructuring together with government rescue packages and a mild regulatory regime. Relieved of heavy debts and government demands for rapid 3G investments in rural areas, the operators could generate just enough cash flow to continue their 3G investments but at a slower pace.

Applications and services focus on convenience for the user

Users keep all their information stored at their favorite big company portal, easily accessible from anywhere, at any time. There are numerous services available, but most users prefer the convenience of a one-stop solution. Wireless devices are used for payments, to get profiled advertisements based on geographical location, secure transactions of money between peers, and so on.

No free airwaves

Governments have been very slow to release new spectrum during the past decade. Unlicensed spectrum use is heavily limited by extremely low upper limits of emitted power. As the only spectrum owners for wireless, mobile operators remain the dominant gatekeepers in the industry.

Trends and Fundamental Drivers

Underlying all four scenarios are a set of fundamental drivers, or "mega-trends", shaping the development of the wireless world. From these drivers we have identified 14 important trends whose direction and rate of change are uncertain. We created the scenarios by assigning different values to each trend (high or low, fast or slow, etc.).

Fundamental drivers

The fundamental drivers (Table 6.1) are a compilation of common wisdom from several broad areas: technology, socioeconomics, politics, business and industry, as well as users and user values. We believe these drivers are valid today and will be valid in 2015.

Trend 1: Development will be more user driven

Up to now vendors and technology have driven the wireless development. This will probably change but it is not clear how much the development will be user driven and which user segments will be the most important drivers.

Trend 2: User mobility will increase

In the future we will probably travel more and for longer and we will spend more time commuting. However, we don't know how fast traveling will increase and how we will travel.

Trend 3: The service and application market will grow

The future market for wireless services will probably be much larger than today, consisting of complex and basic services. We might see an abundance of different services and service types or rather few.

Trend 4: User security, integrity, and privacy will become more important

Guaranteeing security, integrity, and privacy is an important problem facing the industry. The difficulty and complexity of this issue make it impossible to predict whether it will have been solved by 2015.

Trend 5: Real or perceived health problems due to radiation
will become more important

A big threat to the industry is health problems, real or perceived, due to radiation from devices and networks. Research might indicate that the radiation is in fact harmless, but it might show it is dangerous.

Trend 6: Environmental issues will become more important

The trend towards increasing environmental awareness will continue. Two areas of special importance are energy consumption and potentially detrimental substances used in terminal cases and other equipment. It is unclear how large these problems will be in the future.

Trend 7: Spectrum will become an increasingly scarce resource

Today most of the spectrum is locked in by legacy users, e.g., operators, the military, and television broadcasters. The shortage is forcing operators to build unnecessary and expensive infrastructure. Growing usage will aggravate this problem. Regulators might decide to release a lot of spectrum or very little.

Trend 8: The wireless industry will grow

The wireless communications industry will probably grow during the coming decade, both in size and scope. The question is how fast.

Trend 9: The big NICs will continue their positive development

There are many signs of positive development in the most important NICs, e.g., China and India. These telecom markets are very large, they grow rapidly, and new companies are established with an ambition of becoming global players. Their future importance on the wireless scene might be large or small.

Trend 10: Market concentration in the wireless industry will change

The future structure of the wireless industry is an open issue. We might see an increased concentration with a few market leaders wielding great market power or a fragmented marketplace where the market leaders have little power.

Trend 11: The fight for market dominance in the wireless industry will intensify

The merging of telecom, datacom, and media into a single industry will have an important impact on the existing telcos. It is not clear which industry will emerge as the winner.

Trend 12: Short terminal usage time and complexity management will become increasingly important problems

Power consumption in the mobile devices and how to simultaneously manage many complex and heterogeneous wireless systems are two crucial problems. They might be solved by 2015, or they might not.

Trend 13: 3G will be implemented

Currently one of the most important issues for the wireless industry is the deployment of 3G. It seems clear that 3G will be implemented, but at what speed and to what extent?

Trend 14: Protecting IPR on content will become increasingly difficult

The problem of protecting intellectual property rights (IPR), especially on content, is very important for the industry. We don't know whether it will be solved by 2015.

Technological Conclusions from the Scenarios

From the scenarios we have formulated a number of technical implications about the technology used in 2015. Assuming they are true in 2015 means that the underlying problems and bottlenecks we face today have been solved by then. The technical implications can be summarized as follows:

• The wireless infrastructure will be heterogeneous
• Efficient and very high rate air interfaces will be developed
• Traffic will be IP based

- Much of the access infrastructure will be ad hoc deployed
- Cost per transmitted bit will be very small
- There is no harmful radiation from base stations
- The power consumption in the wireless systems will be reduced
- Terminals will have a wide range of shapes and capabilities
- Terminals will be cheap, very small, and modularized
- Usage time without charging the battery will be very long
- User interfaces will be highly developed and advanced
- M2M (machine-to-machine) communication will be very common
- Wireless devices will be harmless to people and the environment
- Wireless services will become a commodity
- Services will be independent of infrastructure and terminals
- Telepresence and emotional communication will be available
- Content will be highly personalized
- Global roaming and seamless services will be possible
- Broadband services will be available for all transportation systems
- The end user will be always best connected
- Ubiquitous computing will be everywhere
- Very high levels of security can be provided

Challenges for Research, Industry, and Key Regions

It is clear by looking at the technical implications that there are several important challenges facing the wireless industry and the research community in the next 10–15 years. These are topics we believe are critical for a positive development and where industry can stumble if things go wrong or are left unresolved. One important conclusion is that the academic research tradition of well-established and very specialized disciplines will not suffice in the future. Expertise from many different fields and from industry and academia will be necessary if we are to solve the most important research problems. Several different engineering disciplines are crucial but equally important are economics and business administration, psychology, sociology, and medicine. Cross-disciplinary and cooperative research is becoming increasingly important.

Low-cost infrastructures and services

The cellular infrastructure we see today has been deployed under a high-cost business model, which has been possible due to high user revenues. This will not be viable in the future. Users are not prepared to see their wireless bills increase by several hundred percent, which is necessary if future wireless multimedia are to be carried over traditional networks. Therefore we need to develop technologies for providing wireless bandwidth at affordable cost in a world of many heterogeneous networks. The problem with conventional cellular systems is that they don't scale in bandwidth in the economic sense. A large part of the infrastructure cost is related to network planning, construction, and site work. Economies of scale, and certainly Moore's law, are not applicable on site acquisition, roadworks, erecting towers, etc. In fact, total cost depends rather weakly on the basic radio technology since current modulation and signal processing technologies are quite advanced and very close to the theoretical limits that not even a radical improvement in processing capabilities will significantly improve performance. Since users are accustomed to being connected anytime and anywhere, these parameters can hardly be compromised. If affordable and high-quality multimedia services are to be possible, radically cheaper network architectures have to be developed. Here are some of important research topics:

- Networks that are very easy and cheap to design, implement, and maintain, such as networks deployed ad hoc
- Decentralized resource management techniques (e.g., spectrum, power, and available infrastructure) for large and complex networks
- Flexible allocation of network capacity in time and space while taking advantage of specific traffic patterns (e.g. asymmetric traffic)
- Dynamic spectrum allocation between different services and systems, perhaps according to fluctuations in user demand over time
- Sharing of infrastructure and spectrum between many operators
- Techniques for improving radio link quality, for example smart antennas and MIMO (multiple-input multiple-output) channels
- Business models where all companies in the value chain get a reasonable share of total revenues

Seamless mobility

In a future with billions of users connected over many heterogeneous networks with a multitude of architectures and air interfaces, complexity will be much higher than today. It will be difficult to provide seamless mobility with uninterrupted services, global roaming, and secure billing. Different systems and access networks have to be integrated into a single network, transparent for data traffic. Here are some important research topics:

- Decentralized (distributed) control and management systems
- Standardization of interfaces on different system levels
- Adaptive and multimode networks and terminals that can switch between different air interfaces and protocols, such as through flexible software radios and modular system design
- Scalability of services and techniques for supporting different QoS levels
- Services based on collecting, memorizing, and organizing information in the terminal in a smart and personalized manner
- Techniques to support user mobility, such as user identification, handover procedures, roaming, and billing

New and advanced services

In the future we will see a multitude of different services, some simple and some very advanced. Computers will gradually move into our surroundings, becoming invisible to us. Being always connected and having continuous access to computational resources will lead to ubiquitous and seamless services. Here are some important research topics:

- Context- and location-aware services including requirements for terminals and infrastructure
- Smart spaces with a multitude of displays and sensors surrounding the user
- Techniques for guaranteeing user security and privacy
- Techniques for delivering services in a scalable manner, for low or high data rates, for small or large screens, for low or high prices, etc.
- Efficient and high-rate air interfaces, where OFDM (orthogonal frequency division multiplexing) and UWB (ultra wide band) are interesting candidates

Usability and human–machine interface

At present, the user interface of wireless terminals is quite primitive, based on buttons, joysticks, roller wheels, or simple voice recognition. Most interfaces rely on proprietary models and are neither standardized nor intuitive for the user. With the very rapid development of processor power and memory capacity, the power consumption of wireless terminals will increase dramatically, especially when advanced services are used. At the same time, battery capacity develops much slower. Research should be focused on the following areas:

- Splitting the mobile device into simple information appliances; each appliance takes care of a specific task and is linked to the other appliances, perhaps using a PAN
- User- and human-centered systems where the technology is integrated in everyday things
- New input and output technology, such as advanced voice recognition and display technology
- Devices and networks for users with special needs, such as failing eyesight or hand tremors
- Techniques for reducing power consumption of mobile devices and increased battery capacity

Health and environment

The effect of electromagnetic radiation on the human body is an area of crucial importance for the wireless community. As yet, no generally accepted scientific research has proved that use of wireless terminals is dangerous, at least not with the radiation levels allowed today. But on the other hand, it has not been proved, and probably cannot be proved, that electromagnetic radiation is completely harmless. This complex problem has to be taken very seriously by the industry. Even if, as many experts argue, the radiation levels permitted today are in fact harmless, radiation remains a threat that needs to be dealt with. The problem is that no proof of danger is not the same thing as proof of harmlessness. If users are afraid, it is a problem, justified or otherwise. There is clearly a need for more research in this area. Another issue is the power consumption in communication networks and devices such as computers, servers, and base stations, and when charging terminals. Research is needed here too.

Threat from disruptive market change

In the coming years we will see a continued integration of the telecommunications, IT, and media industries. Each is a huge and influential industry in its own right. The big question is, What will happen to the traditional telcos, especially equipment vendors, terminal vendors, and operators? In the wireless industry, large resources are spent on research and development. Very large numbers of users have been attracted by wireless services during a relatively short period. The volumes on the handset market are impressive, over 400 million units per year. The wireless systems, in particular the cellular networks, are very large and complex but still reliable. At the same time, they are expensive to plan, build, maintain, and operate. The planning cycles are long; standards are developed and agreed in a slow and complicated process. What will happen when these companies are seriously attacked by competitors from the IT industry with a radically different way of doing business, developing products, and standardizing systems, products, and services? There are technologies and services on the market today that may ultimately threaten the dominance of the big telcos. Most of these originate in the IT industry, for example products and networks based on IP, self-deployed and self-configurable peer-to-peer networks, open APIs, and modularized terminals. Many of these are used to offer communication services in the unlicensed part of the spectrum. These products and technologies don't have the reliability and sophistication of traditional telco products. But what if they are good enough? What if users accept best-effort solutions, if they are offered at a low enough price? The telecommunications industry is in for a serious attack from the datacom world. Who will be the winner is not clear.

Smarter spectrum release

On many wireless markets and especially in urban areas, capacity shortage is a serious problem. Much research and development is focused on increasing spectral efficiency and on finding smart ways of sharing spectrum between users or operators. Even so, shortage of radio spectrum is potentially one of the most serious inhibitors to a fast and positive development of the industry. If regulators, in a smart way, allocate more spectrum for wireless communication, growth will be much faster and prices lower. Still today, most spectrum of interest for these services is controlled by other users, for example educational institutions, the military, and television broadcasters. Public communication services have been given less than one-tenth of all usable spectrum between 0.5 GHz and 5 GHz, causing a shortage in many places and forcing operators to build unnecessary and expensive infrastructure.

3G and the telco debt threat

An all too obvious threat is the enormous debt burdening many wireless operators after the financial hype of the late 1990s and early 2000s. Companies in Europe, the US, and elsewhere have paid enormous amounts for 2G and 3G licenses. In addition, many operators are facing future investments of the same magnitude in order to build the networks. The business case for 3G would be more reasonable if the networks were allowed to grow organically with usage.

All industries mature

The telecommunications industry is old, dating back to the second half of the nineteenth century. It has gone through periods of fast technological development and rapid growth followed by periods of stagnation and slower development. The past decades were a period of very rapid development, especially in wireless communications. Most observers consider wireless a hi-tech industry with a large potential for growth and for changing society and the way we live. But if we look back in history, we see that all industries, even old hi-tech industries, eventually matured and entered a phase with slower technological development, lower profit margins, and different business logics. The question is not if, but when telecom and wireless will mature.

Challenges for the US

Despite its dominant IT industry, the US lags Europe and Japan by two to three years when it comes to wireless services. Here are some key challenges for the US wireless industry:

* Increase the use of wireless data services by improving user quality of service, decreasing price, decreasing roaming charges, and by developing new, attractive, and secure services
* Develop and deploy infrastructure and technologies that increase the poor coverage and alleviate the capacity problems in many urban areas
* Reduce the problem of poor interoperability between services and devices
* Develop solutions and standards to deal with the problems of having several competing 2G and 3G standards
* Consolidate the operator industry to reduce fragmentation and have fewer companies with a very small market share

Challenges for Europe

Europe is one of the most important economic regions of the world, and its importance will increase following enlargement of the European Union. Europe is the world leader in wireless voice services, but uptake of data services has been very slow, especially compared to Japan and Korea. Here are some of the challenges facing Europe:

- European wireless operators have to become more user driven and less technology driven, as shown by the fiasco of introducing Wap
- The debt crisis facing many of the large operators needs to be solved; it is creating almost a freeze on infrastructure investment and this feeds through to the vendors
- The very strong European telecommunications industry might come under serious attack from IT and datacom companies introducing new technologies and products that compete with the traditional telco products
- The possible negative effects on humans of electromagnetic radiation might scare users into avoiding wireless technology and force governments to regulate the business
- In scientific research, Europe has problems competing with the US and perhaps also with Japan
- Europe's relatively strict labor regulations and its rapidly aging population are other areas that need to be addressed in order for Europe not to be left behind by other regions

Challenges for China

As the biggest market for wireless products and services, China has a huge potential for the future. Vendors with global ambitions have already appeared on the scene. However, the country also faces some serious challenges:

- Establishment of new and reformed institutions and macroeconomic transparency will be critical for a positive industrial development in the long run without letting the high growth rate of the Chinese economy distract attention from the need for deregulation and structural reforms
- It must deal with the threat of political instability when a new generation demands more political rights and freedom due to better education, access to more information and technology, and a higher level of material wealth

- If China decides to develop its own 3G standard, the consequences will probably be severe, since the domestic vendors might lose their chance to compete globally and the global vendors will have to develop products adhering to yet another standard in order to compete in the Chinese market
- The Chinese vendors face big challenges to their global ambitions, because the large operators have well-established relations with Western vendors, giving them a significant advantage in terms of handling the complexity of the telecom systems and the need for backward compatibility with the existing systems

Challenges for Japan and South Korea

Japan and South Korea are the world's most prominent markets for wireless data services. They are essentially the only big countries where multimedia services over 2.5G networks have taken off. In Japan the three major operators have launched 3G services, albeit with a somewhat slower uptake than expected. Here are the main challenges facing the wireless industry in Japan and South Korea:

- Translate the successful 2.5G experience into successful 3G services despite stagnating ARPU (average revenue per user), expensive and complex handsets, and short battery life
- Avoid the problems associated with markets where there is one very dominat player and few other players, which is the case in the wireless operator markets of Japan and South Korea
- Terminal vendors and network vendors in Japan and South Korea lack a strong global market position and it will be a tough fight to gain one
- The prolonged Japanese recession and the developments in the conflict on the Korean Peninsula are potential inhibitors to positive development

Moving into the Future

Thinking and speculating about the future is always a problem for anyone living in a media-saturated world. The imminent risk is to extrapolate the current fads and forget other possible perspectives on the world. During the recent telecom bubble all curves pointed upwards and exuberant optimism concealed almost all alternative viewpoints. In the current backlash, pessimism is allowed to dominate. To avoid this trap it is necessary to take several steps back and view the world from the vantage point of decades.

Our scenarios are visions, not predictions, painted to show possible future developments. The four scenarios are not mutually exclusive. They point in four different directions in the 14-dimensional trend space, which is the benefit of avoiding the 2×2 matrix when designing the scenarios. More than one scenario can be valid at the same time. Parts of two, three, and possibly all four scenarios can be valid when looking back at them with hindsight.

There are many aspects and dimensions that we have overlooked in the scenarios. We have not considered discoveries in the science labs after 2010 that will lead to new technologies and products reaching the market after 2020. Nor have we considered the impact of a possible large change in regional leadership. What if the Chinese wireless industry is the global leader in 2015?

It seems that new technologies receive the most attention and enthusiasm at the early stages of their development. This is where everybody knows about the benefits of the new technology but few actually have access to it. The railway was one of the great technological wonders of the nineteenth century. England was the first country to embrace railways and a great railway hype led to a bubble that burst between 1845 and 1847. When the US began to embrace railways a few decades later, it inflated a second large railway bubble that burst in 1873. The investors in the bubbles lost their money but the infrastructure that was already deployed eventually came into use under new owners. The bursting of the bubbles was merely a small dip during the early stage of a major technology rollout. The large impact on society came decades after the bubbles. But then the railway was a commodity taken for granted. There are other examples of hype during an early stage of a new technology: cars in the first decades of the twentieth century, commercial airlines in the 1950s, radio broadcasting in the 1920s, and the first biotech bubble in the early 1980s. Wireless might become another.

One way of modeling how new technologies first create a peak of inflated expectations long before the technology in question reaches its full impact on society is the hype cycle from Gartner Dataquest (Figure 12.1). The hype cycle has been developed to explain the lifecycle of new information technologies. It shows how a very high level of attention at an early stage of development (peak of inflated expectations) is followed by a backlash (trough of disillusionment). Eventually the new technology starts to deliver on its promises (plateau of productivity).

After the burst of the tech bubble in 2000 it is obvious that in 2003 we are on the way down from a peak of inflated expectations. What we don't know

Figure 12.1   The hype cycle for new technologies (Reprinted with permission from
Gartner Dataquest)

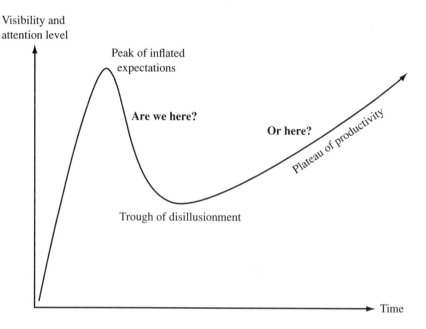

is whether the downturn will continue for years or whether we have already
come past the bottom of the backlash. However, we can probably say that by
2015, IT and wireless technologies will have a much larger impact on society
and business than today and they will be regarded as commodities—they will
be taken for granted. Hence our fundamental view is towards the optimistic
side. In a hundred-year perspective, economic growth and technological
development will push the curves upwards.

Dear Reader in 2015

You might not understand the rationale or the effort behind this book at all.
The paradox of our wireless foresights might be that the more efficiently the
wireless services manage to deliver on their promises, the less people will
notice them. There might not even be wires around in your time!

Appendixes

Appendix A

User Segments

This is a short description of what the main consumer segments might look like in the wireless market in 2015. It is the result of our own analysis and a combination of findings in reports and surveys covering this topic in commercial, governmental, and academic reports. The terms "Moklof" and "Yupplot" are taken from Lindgren, Jedbratt, and Svensson (2002) but our descriptions might be slightly different.

Moklofs

Moklofs are mobile kids with lots of friends. They are often seen as the driving consumer segment, but not necessarily the best in terms of spending money on the services they use. Together with some parts of the Yupplot segment, they are usually seen as the early adopters of new technology. Moklofs are generally young people that have yet to start their own families or careers. They are strongly focused on entertainment and messaging services that require less security than many other services. They participate in communities a lot, both local and global, and are very global in their ways of thinking. Napster-like services are a perfect example, newsgroups another. Both are also good examples of services that are a bit underground, something that the very postmaterialistic Moklofs have embraced quickly. Content is king for this segment, leading to a vast number of different services and applications. Games and music are popular, but social and environmental services have also turned out to be successful.

If this segment turns out to be dominating or leading the way, the field for content and service providers will be wide open. This will also mean that the amount of data traffic will be relatively high (e.g., various kinds of broadcasting service) and we will also see a situation where different media types become integrated and interactive. Moklofs are open-minded about new technologies but they are very sensitive to trends. The technology is not only a way to communicate with your friends, but also a way to indirectly communicate who you are, depending on what phone model you have, and so on. Moklofs trade personal information to get better services and better deals. Integrity issues are not very important, something that might lead to faster and cheaper development of a variety of services. This also mean that brand will be increasingly important; Moklofs will be happy to provide their favorite brands with more information about themselves in order to get better, newer, or cooler services; cheaper fares; and so on.

Yupplots

Yupplots are young urban people/parents with lack of time; they are a more mature segment than the Moklofs. Time and efficiency are keywords for the Yupplots, who try their best to combine career with family and friends. Social issues such as being close to friends and family are becoming increasingly important, yet the answer might not be just messaging services—although they will be important anyway—but services that also save time or make things more efficient. Truly mobile services allowing Yupplots to work while commuting or at home are likely to become successful, as well as services that save time, like banking or buying goods on the move. Other, perhaps more surprising, services might be surveillance services that allow working parents to keep track of their kids at daycare or to check up on other household issues, such as groceries that need to be bought on the way home from work. But in 2002, out of all the user segments, the Yupplots are the most stressed by information technologies.

Yupplots are not very sensitive to technology trends. Two of their main demands are that the technology is easy to use and that it provides them with services and applications which give them more time for their family and friends. Handling business correspondence on the move and updating the office calendar are two obvious services that will become widely used. Others are transaction services such as planning dinner and buying tickets on the move. This means that security and integrity issues are important, something that will influence the market for content and service providers and will probably put considerable emphasis on issues such as brand.

Elders

In 2015 the 1940s generation will be retired; it will be a huge group in society. Regardless of what one might think, this group will be relatively technology-friendly since a large proportion will be well educated and computer literate after a career where computers and other types of information technology have played an important role. Elders demand independence and an active lifestyle. Being old is not the end of one's life, just an opportunity to do all those things one never had time for in earlier years.

Elders place high demands on usability and quality of service, and they are not afraid of letting their voice be heard. Brands are important but not only in a commercial way; they also matter in public services.

Apart from peer-to-peer communication—keeping in touch with family and friends—healthcare services will become a huge market, allowing people to check their health wherever they are. This highlights integrity and privacy, adding to the importance of brand, and it also affects the amount of data being transferred. Videoconferencing might be an important way of communicating with institutions dealing with personal health or community services. Although technology-friendly, this segment is quite sensitive to the health effects that new technologies might impose, real or otherwise.

Mobile Professionals

Mobile Professionals are a very price-insensitive segment requiring high-quality information services as well as commerce and business services and applications. They are usually seen as early adopters of new technology.

Mobile Professionals are very mobile and do a lot of globe traveling. It is important for them to be able to connect to home offices to synchronize work and correspondence, but they also use available time to catch up on the latest news from around the world or plan the coming week. Stock market information is a typical service, together with streaming broadcast news and online ordering of tickets and other goods. Personal communication with family and friends is also important. They use videoconferencing, but almost exclusively as a communication tool for business meetings. Security is a top priority, at least seen from their employer's perspective, together with seamless technologies providing relevant information at the right time. To some extent this segment is using hidden communication between the personal device and the home base, i.e., machine–machine communication. Global coverage, seamless communication, high-quality services, personalized information, security, and high data rates are essential for this segment. The terminals they use are powerful and expensive.

Industrial Users

Despite being a Yupplot or a Mobile Professional, some users have jobs that make them Industrial users more than anything else. Industrial users are people working as sales staff, logistics experts, couriers, etc., jobs that make them highly mobile and in need of fast and high-capacity data links with their offices or organizations. In many cases the links might take the form of hidden communication between the personal terminal and a central server, i.e., machine–machine communication.

This segment is not particularly concerned with ease of use but more concerned with issues such as data speed, coverage, positioning, security, tailor-made services, and integration with other systems. Costs are less important and the terminals might become something between a PDA and a laptop, i.e., rather powerful and expensive.

Appendix B

Wireless Foresight at Wireless@KTH

The Wireless Foresight Project

The scenarios in this book were developed in a project called Wireless Foresight, carried out in 2001 and 2002 at the research center Wireless@KTH. The main objectives for the center when initiating Wireless Foresight were (1) to initiate and drive a strategic discussion on what research and other activities to perform within the center, (2) to create and maintain a shared vision about the center's goals and, (3) to provide visibility.

To identify relevant research topics and assess their importance, it is necessary to create plausible and consistent descriptions of future environments and settings for the technology in question. Wireless Foresight aimed at creating such scenarios. The objectives of the project were (1) to create scenarios for the future of the wireless telecommunication and data communication industries and technologies, including actors, users, and systems, as well as development paths leading there; and (2) to identify key research areas for the center and in general.

The research issues addressed at Wireless@KTH aim quite far into the future. Research is carried out on the wireless systems that will be deployed and used in 2010–2015, after the third-generation cellular systems being introduced today. This is the reason why 2015 was chosen as the time horizon of the scenarios. The scenarios principally focus on the challenges and the development of the wireless industry and its actors. The industry has been divided into operators, infrastructure vendors, terminal vendors, and service providers and developers—the typical industrial partners of Wireless@KTH.

Wireless@KTH and the Vision-Driven Research Approach

Wireless@KTH is a center for research and education on wireless systems created by the Royal Institute of Technology (KTH) in Stockholm in cooperation with industry. The center was inaugurated in spring 2001 with Ericsson and Telia, the largest Swedish telecom operator, as founding partners together with six research groups at KTH. Today a large number of companies, organizations, and a research group from Stockholm School of Economics have joined the center and participate in research projects and other activities.

The goal of Wireless@KTH is to be an internationally leading research facility on wireless communications. The center engages in large inter-disciplinary research projects in collaboration with industry, as well as smaller and more focused efforts. Wireless@KTH also arranges conferences, seminars, executive courses, and develops new courses and curricula for graduate and undergraduate students at KTH and elsewhere.

The Wireless Foresight project played an important role in the process of formulating a clear vision and a long-term strategy for the center. The scenarios and the key research areas constitute the basis of the vision-driven research approach at Wireless@KTH. Through the vision created by the scenarios, research areas to be targeted by the center are identified and can then be linked to specific research projects. The vision of the future is continuously discussed and updated, leading to new or reformulated research areas and projects.

More information about Wireless@KTH and its industrial and academic partners, research activities, partnership program, and the Wireless Foresight project, can be found at www.wireless.kth.se.

Glossary

2G	second-generation mobile telephony, digital voice (GSM, IS-95, AMPS, etc.)
2.5G	2G with added data capacity (GPRS, etc.)
3G	third-generation mobile systems (IMT2000)
4G	a broad concept for advanced mobile technologies that might reach the market after 3G beyond 2010. 4G is a buzzword today and is not used in this book
AAA	authentication, authorization, and accounting, the three functions required for secure transactions
AMPS	American Mobile Phone System, American version of 2G
API	application program interface, the set of functions specified by the software vendor that other programmers can use to communicate with the software when integrating it with their own systems
ARPU	average revenue per user
ATSC	Advanced Television Systems Committee, US standard for digital TV competing with DVB-T
BAN	body area network, network involving only devices on the body
Blackberry	two-way paging service offered over a Mobitex network
BFR	brominated flame retardant

Bluetooth	wireless standard for short-range cable replacement
CDMA	code division multiple access
CDMA2000	a 3G standard competing with WCDMA
codec	coder and decoder
CTO	chief technology officer
DAB	digital audio broadcasting, standard for digital radio
DVB-T	digital video broadcasting, terrestrial, the dominant global standard for digital TV competing with ATSC
DVD	digital versatile disk, used for storing video or other data on a CD-sized disk
EDGE	Enhanced Data for GSM Evolution, an upgrade of 2.5G
Elder	user segment, see page 207
FOMA	Freedom of Multimedia Access
FWA	fixed wireless access
Gb	gigabit, unit for measuring data storage and traffic capacity
GDP	gross domestic product
GPRS	General Packet Radio System, a 2.5G technology
GPS	global positioning system, system for geographical location using triangulation of signals from GPS satellites
GSM	Global System for Mobile, 2G mobile telephony
HAP	high-altitude platform, such as an airship or balloon
hotspot	small location where broadband wireless access is available
iMode	wireless technology for mobile internet in Japan
Industrial user	user segment, see page 208
IP	Internet Protocol
IPR	intellectual property rights
IR	infrared
IS-95	a 2G standard based on CDMA
ISP	internet service provider
ITS	intelligent traffic system
J2ME	Java 2 Platform, Micro Edition
kbps	kilobits per second
LAN	local area network
LEO	low earth orbit
M2M	machine-to-machine
MIMO	multiple-input multiple-output
MMS	multimedia messaging service
Mobile Professional	user segment, see page 207
Mobitex	packed-switched, narrowband, data-only mobile communication network. It is public and based on an open international standard
Moklof	mobile kid with lots of friends, see page 205

MVNO	mobile virtual network operator, a mobile operator without its own infrastructure. The MVNO buys network capacity wholesale from mobile networks and resells it to consumers under its own brand
Napster	example of (former) web service for free sharing of IPR-protected media
NIC	newly industrialized country
OFDM	orthogonal frequency division multiplexing, air interface with high capacity used in WLANs
PAN	personal area network, immediate vicinity around the person
PDA	personal digital assistant
PIN	personal identification number
QoS	quality of service
R&D	research and development
RF	radio frequency
SAR	specific absorbed radiation
SIP	Session Initiation Protocol
SMS	short messaging service
SUV	sport utility vehicle
TDMA	time division multiple access, a coding standard for the air interface used in 2G systems such as GSM
UMTS	Universal Mobile Telecommunication System, alternative name for the 3G standard
UWB	ultra wide band
VoIP	voice over IP
VPN	virtual private network
WAP	Wireless Application Protocol, protocol for GSM and GPRS data traffic
WCDMA	wideband CDMA, a coding standard for the air interface used in UMTS
WiFi	international association to certify interoperability of WLAN products based on the IEEE 802.11 specification
WISP	wireless ISP, also known as public WLAN or P-WLAN
WLAN	wireless LAN
WRC	World Radio Conference, diplomatic body for international settlement of spectrum issues
WWRF	Wireless World Research Forum
Yupplots	young urban people/parents with lack of time, see page 206

References

3 (2002) *Information, Communication, Consumption: Our Lives 2010 and beyond!* 3: Stockholm.

Arvedson, M., Edlund, M., and Thomsson, K. (2002) Implementing Bluetooth internet access through WLANs? A WISP decision criteria evaluation. Student project, Wireless@KTH, Royal Institute of Technology (KTH), Stockholm.

Bergqwist, J., Engren, J., and Eriksson, O. (2002) WLAN as a complement to UMTS: performance comparison, business and market drivers. Student project, Wireless@KTH, Royal Institute of Technology (KTH), Stockholm.

Bhattacharya, S (2002) Telecommunication scenario in India. Working paper, Wireless@KTH, Stockholm. Available at www.wireless.KTH.se/foresight/.

Bria, A., Gessler, F., Queseth, O., Stridh, R., Unbehaum, M., Wu, J., Flament, M., and Zander, J. (2001) 4th generation wireless infrastructures—scenarios and research challenges. *IEEE Special Edition Personal Communication Magazine*, The full PCC report is available at www.s3.KTH.se/radio/4gw/public/Papers/ScenarioReport.pdf.

Christensen, C. (1997) *Innovators Dilemma: When New Technologies Cause Great Firms to Fail.* Boston MA: Harvard Business School Press.

Christensen, C., Verlinden, M., and Westerman, G. (2002) Disruption, disintegration and the dissipation of differentiability. *Industrial and Corporate Change*, **5**, 955–993.

Dahlbom, B. (1999) *Talk society*. Available at www.informatik.gu.se/~dahlbom/.

Dahlbom, B. (2000) Nätverkande nomader. In *Ledmotiv*, no. 3. Centre for Advanced Studies in Leadership, Stockholm School of Economics, Stockholm.

Deutsche Bank (2001) *The Rise of the 3G Empire: Even Rome Wasn't Built in a Day*. Deutsche Bank.

Economist (2001) *A survey of the mobile internet*, October 13. Available at www.economist.com.

Etemad, B. and Lennerfors, T. (2002) Handling spectrum insufficiency in the future wireless world: an academic and industrial approach. Student project, Wireless@KTH, Royal Institute of Technology (KTH), Stockholm.

European Commission (1970–) *Eurobarometer*. Brussels: European Commission.

Grübler, A. (1998) *Technology and Global Change*. Cambridge: Cambridge University Press.

Hofstede, G. (1980) *Culture's Consequences: International Differences in Work-Related Values*. Beverly Hills CA: Sage.

Inglehart, R. (1990) *Culture Shift in Advanced Industrial Society*. Princeton NJ: Princeton University Press.

ISA (2002) Overview of the mobile internet and wireless development in Japan. Unpublished report, Invest in Sweden Agency, Tokyo.

IST (2001a) A vision on systems beyond 3G. Information Society Technologies, Project Cluster.

IST (2001b) Scenarios for ambient intelligence in 2010. Information Society Technologies. Available at www.cordis.lu/ist/istag-reports.htm.

IVA (2000) Swedish technology foresight. Report from Panel 5 of the Royal Swedish Academy of Engineering Sciences, Stockholm.

Lightman, A. (2002) *Brave New Unwired World*. New York: John Wiley & Sons, Inc.

Lind, J. (2002) The 3G backlash—debts and wireless local area networks as the 3G reaper. In *Business Briefing: Wireless Technology 2002*. London: World Markets Research Centre. Available at www.wmrc.com/businessbriefing/technologybriefing/contents/wireless_2002/index.html.

Lind, J. (2003) Structural change during industry life-cycles. Dissertation, Stockholm School of Economics, Stockholm, forthcoming.

Lindgren, M., Jedbratt, J., and Svensson, E. (2002) *Beyond Mobile: People, Communications and Marketing in a Mobilized World*. Basingstoke UK: Palgrave.

Lindmark, S. (2002) Evolution of techno-economic systems—an investigation of the history of mobile communication. Doctoral dissertation, Chalmers University of Technology, Gothenburg.

Long, V. (2002) Chinese market. Working paper, Wireless@KTH, Stockholm. Available at www.wireless.KTH.se/foresight/.

McClelland, S. (ed.) (2002) *Ultimate Telecom Futures: Broadband Multi-Service Networks*. London: Horizon House Publications.

Mishina, K. (1999) Learning by new experiences: revisiting the flying fortress learning curve. In Lamoreaux, N., Raff, D., and Temin, P. (eds) *Learning by Doing in Markets, Firms and Countires*. Chicago IL: University of Chicago Press.

Moore, G. (1991) *Crossing the Chasm: Marketing and Selling Technology Products to Mainstream Customers*. New York: Harper Business.

Moore, G. (1995) *Inside the Tornado: Marketing Strategies from Silicon Valley's Cutting Edge*. New York: Harper Business.

Morgan Stanley (2001) *The Technology Primer 2001*. Available at www.morganstanley.com/techresearch/.

Morgan Stanley (2002) *Global Market Sizing of TMT Products and Services*. Available at www.morganstanley.com/techresearch/.

Nordhaus, W. (1997) Do real-output and real-wage measures capture reality? The history of lighting suggests not. In Bresnahan, T. and Gordon, R. (eds) *The Economics of New Goods*. Chicago IL: University of Chicago Press.

Nordhaus, W. (2001) The progress of computing. Discussion paper, no. 1324, Yale Cowles Foundation, New Haven CT.

Norman, D. (1998) *The Invisible Computer: why good products can fail, the personal computer is so complex, and information appliances are the solution*. Cambridge MA: MIT Press.

Northstream (2002) *Public WLAN services*. Available at www.northstream.se.

Oxford Analytica (2002) China: growth potential. Unpublished briefing, October 10.

Raff D. and Trajtenberg, M. (1997) Quality-adjusted prices for the American automobile industry 1906–1940. In Bresnahan, T. and Gordon, R. (eds) *The Economics of New Goods*. Chicago IL: University of Chicago Press.

Reed, D. (1999) That sneaky exponential—beyond Metcalfe's law to the power of community building. *Context Magazine*, spring. Available from www.contextmag.com.

RISC (1978–) Western scan. Unpublished research, International Research Institute on Social Change.

Sahle, K. and Thuresson, A. (2002) The Russian mobile service market. M.Sc. thesis, Wireless@KTH, Royal Institute of Technology (KTH), Stockholm.

Schildknecht, M. (2002) Market drivers and potential of mobile multimedia services over 2.5G. M.Sc. thesis, Wireless@KTH, Royal Institute of Technology (KTH), Stockholm.

Schwartz, P. (1996) *The Art of the Long View: Planning for the Future in an Uncertain World*. New York: Bantam Doubleday.

Shapiro, C. and Varian, H. (1998) *Information Rules: A Strategic Guide to the Network Economy*. Boston MA: Harvard Business School Press.

SNRV (2002) Mobiltelefoni och hälsoeffekter. Report from Svenska Nationalkommittén för Radiovetenskap, Royal Swedish Academy of Sciences, Stockholm.

Sociovision (1974–) 3SC, socio-cultural monitor. Unpublished research, Sociovision Cofremca, Paris.

Swedish Trade Council (2001) Global perspectives on mobile internet development: a comparative study on 30 Markets. Unpublished paper.

Swedish Trade Council (2002) Wireless opportunities in the United States. Unpublished paper.

WWF (2002) *Sustainability at the Speed of Light*. Stockholm: World Wildlife Fund.

WWRF (2001) *Book of Visions*. Zürich: Wireless World Research Forum. Available at www.wireless-world-research.org/.

Zander, J. (1997) On the cost structure of future wideband wireless access. In *Proceedings of the 47th IEEE Vehicular Technology Conference*, Phoenix AZ, May 4–7, 1997, Vol. 3, pp. 1773–1776.

Author Biographies

Bo Karlson is Wireless@KTH's director of external relations and general manager. He was the manager of the Wireless Foresight project. Karlson holds a Ph.D. in industrial management from the Royal Institute of Technology (KTH). Before joining Wireless@KTH, he was assistant professor in the Department for Industrial Economics and Management at KTH. His areas of expertise include project management, organizational theory, business models, industrial development, and research methodology.

Aurelian Bria is currently pursuing a Ph.D. in the Department of Signals, Sensors and Systems at the Royal Institute of Technology (KTH) in Stockholm. He received his M.Sc. degree in electrical engineering from the Politehnica University of Bucharest, Romania, in 1998. In autumn 2000 he joined the Swedish strategic research program Personal Computing and Communication (PCC), starting his research in the field of wireless infrastructure.

Jonas Lind is a researcher at the Center for Information and Communications Research (CIC) at Stockholm School of Economics, where his research focus is structural changes during the life cycle of the IT and telecom industry. Before rejoining academia, he was a strategy consultant in an internet consulting firm and a senior advisor at Telia headquarters. Lind holds an M.Sc. in engineering and an Econ. Lic. degree in business administration.

Peter Lönnqvist holds an M.Sc. degree in psychology and in 2001 he became a Ph.D. student at the Swedish Graduate School for Human–Machine Interaction. Formerly a member of the Human–Computer Interaction and Language Engineering Laboratory at the Swedish Institute of Computer Science (SICS), he now does research in the design and evaluation of ubiquitous service environments and "the disappearing computer" in the FUSE group at the Department of Computer and Systems Sciences of the IT-University in Kista.

Cristian Norlin holds an M.A. in interaction design from the Royal College of Art in London. He also holds a B.Sc. in multimedia education and technology from Stockholm University. Based in Stockholm, he is working as a consultant focusing on human–computer interaction in areas of concept development, interface design for digital technologies and products, and user-centered development processes.

Index